Parabolic Trough Solar Collectors

Hussein A. Mohammed • Hari B. Vuthaluru
Shaomin Liu

Parabolic Trough Solar Collectors

Thermal and Hydraulic Enhancement
Using Passive Techniques and Nanofluids

 Springer

Hussein A. Mohammed (iD)
Western Australia School of Mines:
Minerals, Energy and Chemical Engineering
Curtin University
Perth, WA, Australia

Hari B. Vuthaluru (iD)
Western Australia School of Mines:
Minerals, Energy and Chemical Engineering
Curtin University
Perth, WA, Australia

Shaomin Liu (iD)
Western Australia School of Mines:
Minerals, Energy and Chemical Engineering
Curtin University
Perth, WA, Australia

ISBN 978-3-031-08703-5   ISBN 978-3-031-08701-1   (eBook)
https://doi.org/10.1007/978-3-031-08701-1

This Springer imprint is published by the registered company Springer Nature Switzerland AG
The registered company address is: Gewerbestrasse 11, 6330 Cham, Switzerland

This book is dedicated to my parents, my wife (Dr. Shayma), my lovely daughters (Danyaa and Lana).

Preface

The subject of parabolic trough solar collectors and the related fields of solar energy technologies have become increasingly important to engineers and technologists in recent years. Research and technology have expanded into a number of diverse areas that have made the study of such topics and devices as PTSC of great interest for present and future potential applications. A thorough study of many of these topics requires, to varying degrees, a description and analysis from the statistical microscopic and macroscopic points of view. In addition, a study of PTSC performance is beneficial in promoting a greater understanding of the foundations, performance analysis, and applications of PTSC. Our intention with this textbook is to provide a self-contained exposition of the fundamentals and applications of PTSC for students, researchers, scientists, and faculties in the engineering thermal sciences.

The key features and contents of the book that readers will hopefully find most valuable are the following:

1. This book will be useful to predict the optical, thermal, exergy, and entropy of PTSC performance for heat exchange processes.
2. This book provides mathematical, numerical, and experimental studies on PTSC thermal performance using passive techniques and nanofluids.
3. This book thoroughly discusses PTSC thermal performance using passive techniques and nanofluids with future research trends of PTSC systems.
4. This book provides a comparative performance investigation of different types of solar collectors and their industrial application.
5. This book discusses industrial operation challenges and scale-up challenges for nanofluid applications in the industrial process.

Perth, WA, Australia

Hussein A. Mohammed
Hari B. Vuthaluru
Shaomin Liu

Acknowledgements

This book could not have been realized and published without the support of a number of people who have directly or indirectly been involved in its preparation process. First, big thanks to my wife, Dr. Shayma, for her unwavering support through every phase of the writing and publication process. I would also like to thank my lovely daughters (Danyaa and Lana) for being such wonderful gifts.

It gives me a deep sense of gratitude to express my sincere thanks to Prof. Hari B. Vuthaluru and Prof. Shaomin Liu who initiated the idea of publishing this book. I am also very grateful for their constant encouragement and continuous support throughout the preparation of this book. It has been a privilege for me to work with amazing people like them. I would also like to convey my sincere gratitude to the authors from around the globe who have produced the relevant papers/journal articles which have been used in the writing of this book.

Introduction

In recent year, parabolic trough solar collectors (PTSCs) have become the most proven industry-scale solar generation technology available today among other concentrating solar collectors. The thermal performance of PTSC is of major interest for optimising the solar field output and increasing the efficiency of power plants. Nanofluids are advanced heat transfer fluids that provide effective thermal and fluid flow characteristics. Nanofluids have been successfully applied not only in heat transfer applications but also in solar energy applications. The utilisation of passive technique (flow turbulators) is also a promising way to solve the low thermal performance and low outlet temperature of PTSC systems. The synergistic effects and detailed study of these recent advances in nanofluids and the utilisation of passive techniques on the PTSC's thermal and thermodynamic performance are of paramount importance for students, researchers, faculties and scientists working in the field of solar energy and nanotechnology processes, thermal management systems, and other engineering-related studies. *Parabolic Trough Solar Collectors: Thermal and Hydraulic Enhancement Using Passive Techniques and Nanofluids* by Dr. Hussein A. Mohammed, A/Prof. Hari B. Vuthaluru and Prof. Shaomin Liu is such a valuable book. Thus, it gives me great pleasure to contribute this foreword.

Before writing this foreword, I have reviewed the book chapters and its details. I was impressed by the book's structure and contents. The book is divided into six chapters. The first chapter introduces different types of non-concentrating solar collectors (NCCs), concentrating solar collectors (CSCs), PTSC's historical start, and available commercial collectors including the currents state of the art. The second chapter focuses on the background of PTSC, basics of PTSC and its performance including optical analysis, thermal performance, exergetic performance and entropy analysis. The third chapter discusses the recently utilised PTSC performance enhancement technique and in particular focuses on the numerical and experimental studies passive techniques with the utilisation of conventional working fluids including the currents state of the art in the field. The fourth chapter discusses the recently utilised PTSC performance enhancement numerical and experimental studies using nanofluids, hybrid nanofluids and other studies/techniques including the currents state of the art in the field. The fifth chapter discusses the heat transfer enhancement methods used in the PTSC-related studies. The sixth chapter summarizes the major conclusions and gives recommendations for future work in the PTSC area of research.

All important data supported by the appropriate figures and tables for better understanding of concepts are presented in this book. This became one of the most important features of this book when I began reading its contents. Thus, I believe this is a very important book for Springer Nature, and it would be very beneficial for students, researchers, scientists and faculties for conceptual understanding of this area of research.

Perth, WA, Australia   Hussein A. Mohammed
                                                                                       Hari B. Vuthaluru
                                                                                            Shaomin Liu

Contents

List of Symbols

A	Area, m^2
a	Aperture width, m
A_a	Collector's projected aperture area, m^2
A_c	Collector area, m^2
A_r	Projected absorber tube area, m^2
Be	Bejan number
C_p	Specific heat capacity, J/kg.K
CR	Concentration ratio
D	Diameter, m
d_{np}	Nanoparticle diameter, nm
d_{ri}	Absorber tube's inner diameter, m
d_{ro}	Absorber tube's outer diameter, m
Ex_a	Solar radiation exergy rate absorbed, W/m^2
Ex_u	Absorber tube exergy rate, W/m^2
f	Darcy friction factor
h	Heat transfer coefficient, $W/m^2.K$
I	Radiation intensity
I_b	Direct solar radiation, W/m^2
k	Thermal conductivity, W/m.K.
$\boldsymbol{k}$	Turbulent kinetic energy, m^2/s^2
$K(\theta)$	Frequency edge modifier
L	Length of the absorber, m
$\dot{m}$	Mass flow rate, kg/s
M	Molecular weight, g/mol
N	Avogadro number, mol^{-1}
Ns	Entropy generation ratio
Nu	Nusselt number
P	Pressure, Pa
P_h	Wetted perimeter, m
P_p	Pumping power, W
Pr	Prandtl number

PEC	Performance evaluation criterion factor
q	Heat transfer rate per unit length, W/m
q''	Heat flux, W/m^2
Q_{loss}	Heat loss per meter, W/m
Q_u	Useful energy transferred to the heat transfer fluid, W/m^2
Re	Reynolds number
r_r	Rim radius (m)
S	Modules of the mean rate-of-strain tensor, 1/s
S_{gen}	Entropy generation rate, W/K
S_{gen}^F	Entropy generation rate due to fluid friction, W/K
S_{gen}^H	Entropy generation rate due to heat transfer, W/K
S_{gen}^T	Total entropy generation rate, W/K
T	Temperature, K
$\overline{T}$	Time-averaged temperature, K
t	Time, s
w_a	Collector's aperture width, m
x, y, z	Cartesian coordinates
ΔP	Pressure drop, Pa

Greek Letters

α_{abs}	Absorber tube absorptivity
ε	Turbulent dissipation rate, m^2/s^3
$\boldsymbol{\varepsilon}$	Emissivity
ΔT	Circumferential temperature difference, K
θ	Angle of absorber tube circumference, degrees
α, Γ	Thermal diffusivity, m^2/s
α_t, Γ_t	Turbulent thermal diffusivity, m^2/s
β	Fraction of the liquid volume which travels with a particle
γ	Capture factor of the mirror and HCE cooperation
Γ	End-loss factor
σ, K	Stefan-Boltzmann constant, W/m^2.K^4
ϕ	Nanoparticle volume fraction
ρ	Density, kg/m^3
μ	Viscosity, Pa.s
μ_t	Eddy viscosity, Pa.s
τ	Transmittance of the glass envelope
τ_w	Shear stress, N/m^2
υ	Kinematic viscosity, m^2/s
η_{el}	Electrical efficiency, %
η_{en}	Energetic efficiency, %

η_{ex}	Exergetic efficiency, %
η_o	Optical efficiency, %
η_{th}	Thermal efficiency, %
φ_r	Rim angle

Subscripts

Abs	Absorber tube
amb	Ambient state
av	Average
b	Bulk fluid state
bf	Base fluid
dp	Dew point
eff	Effective
f	Fluid
f	Focal length (m)
hnf	Hybrid nanofluid
i	Initial
in	Inlet
m	Mixture
nf	Nanofluid
out	Outlet
ri	Absorber tube inner wall
ro	Absorber tube outer wall
o	Plain or smooth PTSC tube
p	Particle
t	Turbulent

Superscripts

-	Time averaged value
′	Fluctuation from mean value

Acronyms

Ag	Silver
Al_2O_3	Aluminium oxide
CFD	Computational fluid dynamics

CPC	Compound parabolic collector
CRS	Central receiver systems
CSP	Concentrated solar power
CTC	Cylindrical trough collector
CTD	Circumferential temperature difference
CNT	Carbon nanotubes
CSP	Concentrated solar power
CuO	Copper oxide
DHW	Domestic heating water
DNI	Direct normal irradiance
DSG	Direct steam generation
EES	Engineering equation solver
ETC	Evacuated tube collector
FLC	Fresnel lens collector
FPC	Flat plate collector
FEM	Finite element method
Fe_2O_3	Ferric oxide
FVM	Finite volume method
GO	Graphene oxide
HCE	Heat collection element
HFC	Heliostat filed reflector
HTF	Heat transfer fluid
IPH	Industrial process heat
LFC	Linear Fresnel collectors
LFR	Linear Fresnel reflector
$MCRT$	Monte Carlo ray tracing
MgO	Magnesium oxide
$MWCNT$	Multiwall carbon nanotube
$NUHF$	Non-uniform heat flux
PDC	Parabolic dish collector
PDR	Parabolic dish reflector
PEC	Performance evaluation criterion
PTR	Parabolic trough receiver
$PTSC$	Parabolic trough solar collector
SBR	Spherical bowl reflector
$SEGS$	Solar energy generating systems
SPT	Solar power tower
SiC	Silicon carbide
SiO_2	Silicon dioxide
$SWCNT$	Single wall carbon nanotube
TiO_2	Titanium dioxide
UHF	Uniform heat flux
ZnO	Zinc oxide

Chapter 1
Background

1.1 Introduction

The sun radiation has high-temperature, and high-exergy energy source, where the amount of its irradiance is approximately about 63 MW/m². Notwithstanding, its configuration drastically diminishes the sun-oriented energy on the Earth's surface down to approximately about 1 kW/m² (Romero et al., 2004). In any case, under high sunlight flux-based motion, this problem can be solved by utilising solar systems of concentrating type, which change sun-oriented energy into thermal energy. Sun-based radiation can be changed over into useful thermal energy by solar thermal and electrical processes respectively. Solar concentrating collectors or flat plate collectors are good examples of sun-powered thermal processes, which could be used to transform sun-oriented radiation into valuable heat.

Many technologies are utilised in industry to convert solar energy into thermal and electrical energies. These technologies are classified as shown in Fig. 1.1. Solar technologies can be classified as either passive or active, depending on the type of system. Active technologies require external components, such as pumps or electronic control, to convert energy. The type of energy supplied to the process is the energy produced directly from the solar collection system used in the industrial process, which can be either thermal or nonthermal (e.g., PV). Solar collector type is determined based on the total radiation transmitted to the reception area compared to the radiation received in the collection area. Solar concentration is obtained by reflecting solar radiation from a large collection area to a smaller receiver area using mirrors or lenses. The energy conversion can be achieved by concentrating technologies using external components, which are usually used to transport a fluid or solar tracking.

Solar concentrating collector systems can be classified based on centre configuration (point-centre concentrators) such as central receiver systems (CRS) and parabolic dishes (PD), where the collector packs sunlight-based radiation in a point

© The Author(s), under exclusive license to Springer Nature Switzerland AG 2023
H. A. Mohammed et al., *Parabolic Trough Solar Collectors*,
https://doi.org/10.1007/978-3-031-08701-1_1

picture to recipient tube, or based on line-centre concentrators such as parabolic-trough collectors (PTCs) and linear Fresnel collectors (LFC), where the collector amasses sun-powered radiation in a direct picture to recipient tube as appeared in Fig. 1.2.

The sun-based radiation is directed onto a central line on the collector axis of PTCs. The liquid streaming inside the recipient tube, which is introduced in the

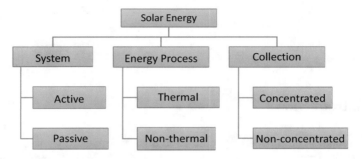

Fig. 1.1 Classification of solar energy technologies

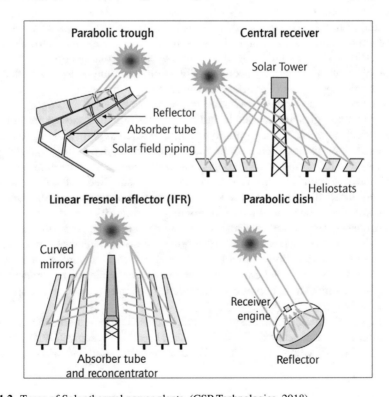

Fig. 1.2 Types of Solar thermal power plants. (CSP Technologies, 2018)

central line, absorbs the concentrated sun-powered energy from the pipe walls and increases its enthalpy. The collector has one-hub sun-oriented tracking to guarantee that the sunlight-based bar falls parallel to its pivot. PTCs usually utilise direct sun-oriented radiation, named bar radiation or Direct Normal Irradiance (DNI), i.e., which is the portion of sun-powered radiation that arrives as a parallel bar at the Earth's surface.

There are two fundamental groups of PTC applications: (i) require a temperature range from 300 to 400 °C, and (ii) require a temperature somewhere in the range of 100 and 250 °C. The first group is associated with Concentrated Solar Power (CSP) plants, which have been effectively tried under genuine working conditions. CSPs have typical aperture widths and total lengths of around 6 m and 100–150 m, respectively, and geometrical proportions are somewhere in the range of 20 and 30. CSP plants combined with PTCs are associated directly and indirectly with steam power cycles. Although the most acclaimed case of CSP plants is the Solar Energy Generating Systems (SEGS) plants in the United States and it is still a work in progress or development around the world. The second group of utilisations are associated with industrial process heat (IPH), heating water for domestic use (DHW), heating, refrigeration and cooling. The second group has standard aperture widths somewhere in the range of 1 and 3 m, complete lengths shift somewhere in the range of 2 and 10 m and geometrical proportions are somewhere in the range of 15 and 20. Most of these facilities are situated in the United States, and other facilities have lately been developed in different nations. These facilities could also be used for different applications such as siphoning water system, desalination and detoxification.

The accuracy and confidence of the predictive theoretical and numerical models are basically dependent on different factors such as applied method type, method structure, geometrical parameters, operational conditions, utilised inputs, etc. Due to this fact, it is crucial to comprehensively investigate the parameters affecting the effectiveness of the suggested methods. In the previous studies, a set of affecting factors are considered and investigated while it is crucial to consider all the influential factors such as the required inputs for each case, structure of the model, employed functions in each model, etc. In this regard, a comprehensive overview is performed on the thermal and hydraulic enhancement utilising passive techniques (and modification in geometries) and various types of conventional and advanced working fluids (nanofluids and hybrid nanofluids) in PTSC. There are some review researches concerning the applications of PTC using classical fluids and nanofluids (Kalogirou, 2004; Jebasingh & Joselin Herbert, 2016; Yilmaz & Mwesigye, 2018). However, in the current study, there is a special focus on the theoretical, experimental and numerical investigations to date in detail including the important aspect of thermal energetic and exergetic analyses and their relations to thermal and exergetic efficiencies of PTC systems. To provide deep insight to the important perspectives that ought to be thought of in PTSC future turns of events, and to facilitate means for researchers to study this area, several scientific sources concerning the prediction of these performances using different kinds of fluids and gases, are reviewed and their most important outcomes are represented. In addition, according to the results of the reviewed articles, some of the challenges and future prospects for the utilisation of nanofluids and hybrid nanofluids in PTSC systems are also addressed in this review paper including the future directions for research. Moreover, since the present study reflects the results of various researchers, it would be possible to select the approach and inputs for proposing more accurate models in the future. The reviewed studies are categorised based on the investigated passive technique used and on the working fluid type (conventional, nanofluids and hybrid types).

The structure of this chapter includes overviews on PTSC types including the general sunlight-based thermal collector types, concentrating and non-concentrating solar collectors. The direct solar absorption collector's efficiency improvement is also discussed.

1.2   Sunlight-Based Thermal Collector Types

1.2.1   Sunlight-Based Collectors

A few sorts of sun-based collectors are available at present accessible in the commercial centre. It can be characterised by their sun-based fixation proportion and following sun's movement (Tian & Zhao, 2013; Thirugnanasambandam et al., 2010). Non-concentrating collectors (NCC) possess a fixation proportion of around one and work with a temperature range from 30 to 240 °C. Regularly, NCC are for all times maintained at a location and do not follow the sun and it depends on their initial direction. In this manner, in the northern half of the globe they face south, and they face north in the southern side of the equator. Concentrating sunlight-based collectors (CSC) are intended to follow the sun's situation during the daytime and can utilise either one-hub or two-pivot following designs. Focus proportions are in the range of 10–1500, possess large radiation motions and can accomplish high temperatures up to 2000 °C as found in Table 1.1. The next sections examine different sorts of collectors as of now being used the world over.

1.2.2   Non-concentrating Solar Collectors (NCC)

The flat plate collector (FPC) is the most widely recognised NCC as appeared in Fig. 1.3. This kind of collector is broadly utilised for water-warming and space-warming applications in homes (Kainth & Sharma, 2014). FPCs are explicitly intended to work in warm and radiant atmospheres. FPCs are rooftop placed, directed and maintained in place for perfect sun-powered presentation. Regularly, a FPC comprises of a protected box fitted with a high transmission properties comparable metal (Hellstrom et al., 2003). There is an absorber inside the box, which comprises of sheets and pipes dealt with a dull shaded sun oriented specific

Table 1.1   Sunlight-based thermal collector types

Collector's Motion		Collector's Configuration	Absorber Type	Concentration Ratio (CR)	Temperature Range (°C)	Reference
Stationary		Flat plate collector (FPC)	Flat	CR ≤ 1	30 ≤ T ≤ 80	Sukhatme and Sukhatme (1996)
		Evacuated tube collector (ETC)	Flat	CR ≤ 1	50 ≤ T ≤ 230	
		Compound parabolic collector (CPC)	Tubular	1 ≤ CR ≤ 5 5 ≤ CR ≤ 15	60 ≤ T ≤ 240 60 ≤ T ≤ 290	
Sun tracking	Single hub	Fresnel lens collector (FLC)	Tubular	10 ≤ CR ≤ 40	60 ≤ T ≤ 270	Rabl (1976)
		Parabolic trough collector (PTC)		15 ≤ CR ≤ 45	60 ≤ T ≤ 400	
		Cylindrical trough collector (CTC)		10 ≤ CR ≤ 50	60 ≤ T ≤ 400	
	Two hubs	Spherical bowl reflector (SBR)	Point	100 ≤ CR ≤ 300	70 ≤ T ≤ 700	Zhang et al. (2013)
		Parabolic dish reflector (PDR)		100 ≤ CR ≤ 1000	100 ≤ T ≤ 900	
		Heliostat filed reflector (HFC)		100 ≤ CR ≤ 1500	150 ≤ T ≤ 2000	

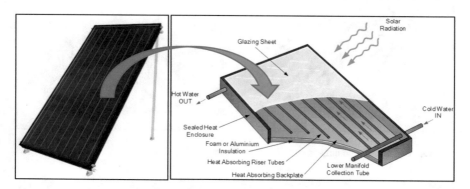

Fig. 1.3   Flat-plate collector (FPC). (CSP Technologies, 2018)

covering to advance sun-based absorption (Konttinen et al., 2003). A large portion of the sunlight-based energy going to the FPC is consumed by the sheet-pipe gathering. The subsequent thermal energy will be transferred to working fluid inside the pipes. The FPC's working liquid (normally water) temperature reaches 80 °C, with this advantage they become perfect for local boiling water (Sukhatme & Sukhatme, 1996). In any case, the operational performance of FPCs is fundamentally decreased during unfavourable climate occasions. For instance, significant lots of chilly, shady and blustery days can incredibly lessen the energy converter efficacy which require additional electrical or gas heating.

The evacuated tube collector (ETC) is an elective kind of NCC as appeared in Fig. 1.4. An ETC comprises of a focal liquid channel contained inside a vacuum-fixed tube assembly. These tubes comprise of two coaxial pipes made of glass, with empty space between the two pipes to diminish the convection and conduction heat losses from the inner pipe that has the flowing liquid (Vasiliev, 2005). This advantage of reducing the convection currents by the vacuum envelope makes the ETC to work at higher temperatures than the FPCs. The external surface of the focal heated pipes is coated with special coatings to absorb much amount of energy, which will be then transmitted to the working liquid (Alghoul et al., 2005). The heating pipes are associated with a complex framework that allows the flow of the working liquid along these lines to FPCs.

Another type of sunlight-based collector is the compound parabolic collector (CPC) as appeared in Fig. 1.5. This kind of collector comprises of numerous inward reflecting parabolic surfaces that immediate sun-powered energy to a receiver pipe situated at the base of the system (Rabl, 1976). The CPCs are equipped for getting a lot of diffuse radiation with no need to follow the sun. The cylindrical tube shape is another type of solar-based receiver. This collector comprises of a round and

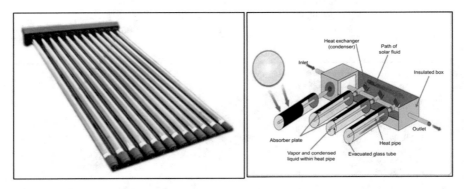

Fig. 1.4 Evacuated tube collector (ETC). (CSP Technologies, 2018)

Fig. 1.5 Compound parabolic collector (CPC). (CSP Technologies, 2018)

hollow tube made of glass that structures the sun-oriented energy receiver and a cylinder that conveys the working liquid. To decrease the reflectivity and advance the transmission of sun-based energy to the focal cylinder, the external glass tube is similar to the available one commercially. The barrel glass tube is made empty to keep away from vapor that develops on the inner surfaces which would lessen the energy transfer. The outside of the cylinder is covered with a particular material that advances greatest sun-absorption, which advances higher temperatures in the working liquid. This working liquid stream rate is managed by valve control to get the greatest energy move (Al-Madani, 2006).

The low thermal energy transformation is one of the disadvantages of using NCC, which in turn limits their utilisation in particular applications that require high temperature. Other disadvantages associated with these types of collectors are its low performance as they experience low fallen sunlight-based energy catch, wasteful heat transfer and low transport properties of the working liquid (Al-Madani, 2006; Minardi & Chuang, 1975). Moreover, receivers are covered with optically specific layer such as dark chromium to improve absorbance but it represents an ecological peril material (Grasse et al., 1991).

Conversely, CSC has the ability to accomplish high thermal energy moves which could be used in power generation system. However, these CSC arrangements are costly and require a lot of money if it is applied in an enormous scale. Moreover, during late seventies, significant efforts were made with the aim to design an alternative collector plans to overcome the shortcomings of the available collectors which initiated the advancement of the direct sun-powered absorption collector. This was done to deliver huge energy moves utilising customary sun-powered collector designs (Lee & Sharma, 2007). However, many variables and components limit the productivity of such sun-oriented collectors such as the sort and number of covers used, surface properties and size of the collector, and collector's materials type.

The convective heat flow rate Q is a significant factor in deciding the effective activity of a collector, which must be considered when planning a sun-oriented collector, and it can be assessed utilising this equation:

$$Q = h\,A\Delta T \tag{1.1}$$

Where h is the heat transfer factor (W/m^2.K), A is the surface area (m^2) and ΔT is the driving force for the heat flow (K) (Eastop & McConkey, 1977). By looking at Eq. (1.1), it uncovers that most of the parameters can be controlled to improve the collector's effectiveness. One of the options is expanding the surface area (A) which looks not practical because the collector's size turns out to be enormous and massive. Another option is expanding ΔT, which requires enhancing the sun-powered concentration proportion, which may not be conceivable because of collector's structure. The third option, which appears reasonable and practical, is to change the heat transfer coefficient by using various working mediums, which have different thermal properties, which will ultimately evaluate the collector's effectiveness.

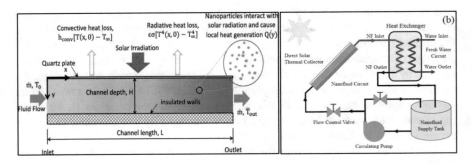

Fig. 1.6 Solar thermal absorption collector: (**a**) Collector's illumination and different heat waste; and (**b**) Closed-loop circuit to transfer heat from nanofluid circuit to water circuit. (CSP Technologies, 2018)

Figure 1.6a shows a schematic of a nanofluid based direct sun-based thermal absorption collector and demonstrates collector's illumination and different heat waste. While Fig. 1.6b shows a closed-loop circuit of direct sun-based thermal absorption structure.

1.2.3   Concentrating Sun-Based Collectors (CSC)

Concentrating sun-based collectors consolidate sun following innovation to keep up the conveyance of focused and centred sunlight-based radiation to the collector. There are four fundamental kinds of CSC: (i) Fresnel linear collectors (FLC); (ii) parabolic trough collectors (PTC); (iii) parabolic dishes (PD) and (iv) heliostats collectors (HC). The FLC can be comprised of either little parabola profile long bended mirrors or from straight sun-powered trackers that have flat reflective strips placed on. The mirrors orient the sun-based light to an optional reflector that focuses and centres the energy onto an absorber pipe. The pipe contains a circulating liquid with temperature in the range of 60 and 250 °C (Gouthamraj et al., 2013). Regularly, FLCs have focus proportions extending in the range from 10 to 40, whereas its direction is accomplished by a single-hub following of the sun's situation during the daytime (Kalogirou, 2004).

The second kind is the PTCs which have been widely examined and its focus proportions is the range of 15 and 45 (Bakos, 2006). The trough is a parabolic or tube-shaped mirror that reflects and focus the sun-oriented illumination onto a pipe. The PTC consists if concentric pipes one is the inner pipe, that has a circulating liquid, made of a dark metal and the outer pipe is made of a glass. Ordinarily, PTCs are fixed on movable sun-oriented trackers that utilises one-hub sun-following frameworks to pursue the sun's movement. PTCs usually work at the temperatures range of 50–400 °C (Kalogirou, 2012).

The third type is the PD, which is generally organised to frame a cluster, where most of the mirrors are coordinated towards a typical point of convergence. PD is usually fixed with two-pivot sun-following frameworks that keep up the focal point of the sun-powered light on the receiver's pipe to pursue the sun's movement during the daytime. PDs has a fixation proportion in the range of 100 and 1000, and it can work at higher temperature of up to 1500 °C. This advantage makes the working liquid can convey enormous amounts of thermal energy to run electrical power generation plant (Nixon et al., 2010). The sunlight-based energy that is generated at the receiver's pipe is then changed over to thermal energy, which is then moved to a reasonable flowing working liquid (Zhang et al., 2013).

The fourth type is the HC, which has organised mirrors to a cluster shape, and these mirrors are driven by two-pivot sun following hardware that empowers the mirrors to reflect and centre the sun-oriented energy into a typical tower (Wei et al., 2010). HC has a fixation proportion up to 1500 and it can operate at higher temperature of up to 2000 °C. HFC systems can create steam with high temperature and high pressure that used to generate power when it is combined with steam generation system (Behar et al., 2013). The different types of CSC are displayed in Fig. 1.7.

The main technical and operational characteristics of the above-mentioned four technologies are given in Table 1.2. It can be clearly seen that the PTC and HC have some advantages, such as maturity and typical capacity, making them the most important technologies used for solar energy conversion. Therefore, PTC is the most deployed technology and has major share in the worldwide energy market compared to other solar concentrating technologies (International Renewable Energy Agency, 2015).

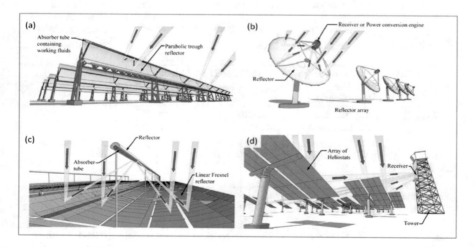

Fig. 1.7 Different types of CSC styles: (**a**) PTC; (**b**) PD; (**c**) FLC and (**d**) HC. (Kalogirou, 2004; Zhang et al., 2013; Nixon et al., 2010)

Table 1.2 Comparison of solar concentrating technologies (International Renewable Energy Agency, 2012, 2015)

Details	PTC	HC	FLC	PD
Typical capacity (MW)	10–300	10–200	10–200	0.01–0.02
Maturity	Commercially proven	Commercially proven	Recent commercial project	Demonstration projects
Technology development risk	Low	Medium	Medium	Medium
Operating temperature (°C)	350–400	250–565	250–350	550–750
Plant peak efficiency (%)	14–20	23-35[a]	18	30
Annual solar to electricity efficiency (%)	11–16	7–20	13	12–25
Annual capacity factor (%)	25–28 (no TES) 29–43 (7 h TES)	55 (10 h TES)	22–24	25–28
Concentration factor	10–80	>1000	>60	Up to 10,000
Receiver/ absorber	Absorber attached to collector, moves with collector, complex design	External surface or cavity receiver, fixed	Fixed absorber, no evacuation secondary reflector	Absorber attached to collector, moves with collector
Storage system	Indirect two-tank molten salt at 380 °C ($dT = 100$ K)	Direct two-tank molten salt at 550 °C ($dT = 300$ K)	Short-term pressurised steam storage (<10 min)	No storage for Stirling dish, chemical storage under development
Hybridisation	Yes, and direct	Yes	Yes, direct (steam boiler)	Not planned
Grid stability	Medium to high (TES or hybridisation)	High (large TES)	Medium (back-up firing possible)	Low
Cycle	Superheated Rankine steam cycle	Superheated Rankine steam cycle	Saturated Rankine steam cycle	Stirling
Steam conditions (°C/ bar)	380–540/100	540/100–160	260/50	N/A
Maximum slope of solar field (%)	< 1–2	< 2–4	< 4	10 or more
Water requirement (m³/MW h)	3 (wet cooling)	2–3 (wet cooling)	3 (wet cooling)	0.05–0.1 (mirror washing)

(continued)

Table 1.2   (continued)

Details	PTC	HC	FLC	PD
Application type	On-grid	On-grid	On-grid	On-grid/off-grid
Suitability for air cooling	Low to good	Good	Low	Best
Storage with molten salt	Commercially available	Commercially available	Possible, but not proven	Possible but not prove

1.2.4   Direct Solar Absorption Collectors Efficiency Improvement

Studies have demonstrated expanding the heat transfer coefficient can fundamentally improve working liquid execution. In any case, available working liquids used to cool down the collectors, for example, water, ethylene glycol and oils have inherently low thermal conductivity than metals as found in Table 1.3. Because of solids have better thermal conductivity than working liquids, it would be a powerful strategy to include limited quantities of strong particles for improving the liquids thermal conductivity. Late investigations have affirmed suspending little particles in a liquid can enhance its thermal performance (Sureshkumar et al., 2013). Specifically, the expansion of little micrometre-scale particles to a working liquid would enhance its thermal conductivity and then increase the heat transfer coefficient (Duncan & Peterson, 1994). Lamentably, two-phase fluids with micrometre-scale particles would have various operational issues in reducing the heat transfer rate such as: (i) molecule sedimentation; (ii) large disintegration rates brought about by circling particles; (iii) particles will in general gather and square tight stream channels; and (iv) expanded stream obstruction and higher pressure penalty (Das et al., 2006). Thus, the utilisation of micrometre-scale particles and nanotechnology-based strategies in practical collectors has revived enthusiasm into creating two-phase liquids.

The capacity to produce nanometre-scale materials with thermophysical, chemical and optical properties that are not quite the same as their mass reciprocals has made the chance to build up another type of liquids named as "nanofluids". It is colloidal suspensions consisting of scattered nanometre-scale particles including metallic, non-metallic, carbides and nitrides. It has a wide assortment of morphologies that incorporate circles, filaments and cylinders (Godson et al., 2010). To improve the direct sun-powered absorption collectors' efficiency, it is believed that nanofluids which have low viscosities, high thermal conductivities and predominant photo-thermal properties are very good option (Saidur et al., 2011). Another issue with nanofluids preparation is its stability, which needs to be considered as well (Hwang et al., 2007). That is why a few examinations have detailed the utilisation of nanofluids and their capacity to enhance the effectiveness of direct sunlight-based energy collectors (Shou et al., 2009).

Table 1.3 Thermal conductivities of different particles and fluids at 25 °C

Particles/fluids	Thermal conductivity (W/m.K.)	Source
Metal		
Gold	315	Perry and Green (1997)
Silver	424	
Copper	398	
Aluminum	273	
Iron	80	
Steel	46	Alghoul (2005)
Stainless steel	16	
Metal Oxide		
Alumina (Al_2O_3)	40	Hwang et al. (2007)
Cupric Oxide (CuO)	77	_____
Iron (II, III)	7	
Titanium Dioxide (TiO_2)	8.37	Kim et al. (2007)
Silicon Dioxide (SiO_2)	1.2	
Zinc Oxide (ZnO)	29	
Carbons		
Amorphous Carbon	1.59	Dean (1992)
Diamond	900–2320	
Carbon Nanofibers	13	
Carbon nanotubes	2000	
C_{60}–C_{70} fullerenes	0.4	
Graphite	2000	
Base fluids		

Particles/fluids	Thermal conductivity (W/m.K.)	Source
Water	0.608	Perry and Green (1997)
Ethylene glycol	0.257	
Glycerin	0.286	
Engine oil	0.145	

Chapter 2
Parabolic Trough Collector (PTC)

2.1 Historical Start of PTC

In 1870, an effective engineer, John Ericsson, a Swedish worker in the United States, structured and manufactured a 3.25 m² aperture collector, which drove a little 373 W motor. He generated the steam inside the sun-powered collector. Then for the period from 1872 to 1875, he managed to manufacture seven comparative frameworks using air as the medium liquid (Pytlinski, 1978). Later in 1883, Ericsson has built a huge "sun engine" in New York. It was comprised of a 3.35 m long, 4.88 m wide PTC, focused sunlight-based radiation on a 15.88 cm width evaporator tube. He designed the concentrator from straight wooden fights, upheld by bended iron ribs joined to the sides of the trough. The reflecting plates, made of window glass, were affixed on these fights (Ericsson, 1884). After 3 years in 1886, he then tried different things with a 1.86 kW sun-oriented motor (Pytlinski, 1978). His "sun engine" was not completed as he passed away in 1889 and his venture was rarely proceeded. Later in 1907, both German engineers Wilhelm Maier and Adolf Remshardt have licensed a PTC with DSG (Maier & Remshardt, 1907).

For the period from 1906 to 1911, Frank Shuman, an American specialist who fabricated and tried various sunlight-based motors. He utilised various sorts of non- and low-concentrating sun-powered collectors having flat reflector wings. These collectors were utilised for siphoning water system in Pennsylvania (United States). In 1912, Shuman structured and introduced an enormous water system siphoning plant in a little farming town (Meadi) located south of El Cairo and close to the Nile River (Egypt). Shuman worked with Charles Vernon Boys, an English specialist, who recommended significant developments of the collectors. Glass-secured kettle cylinders were set along the central hub of a PTC. These sunlight-based collectors created 0.1 MPa saturated steam inside the absorber tube. Each PTC line has 62.17 m in length and 4.1 m width, giving a surface area of 1250 m² and involving 4047 m² of land. The diameter of the absorber pipes was 8.9 cm and had a focus proportion of 4.6, which brought a peak efficiency of 40.7% (Pytlinski, 1978; Kreider & Kreith, 1981).

Shuman and Boys (1917) reflector comprised of various portions of flat mirror with limited spaces between the neighbouring edges. The absorber cylinders and mirror framework were upheld on a lightweight bow formed grid were tilted by methods for a rack and pinion gear. A thermopile or indoor regulator was put in the focal point of the parabola to consequently adjust the absorber to the sun. The collector was not working for whatever length of time it stayed in the shadow of the cylinder.

The Meadi plant was initially appraised at 75 kW mechanical energy and, it gives an account of genuine yield shifted from a little more than 14 kW to a limit of 54 kW (Pytlinski, 1978). It was recommended that with a decent steam motor, the plant could have conveyed nearly 41 kW (Spencer, 1989). Notwithstanding every one of these challenges, the framework was protected in 1917 (Shuman & Boys, 1917). In 1936, Abbot changed over sunlight-based energy into mechanical power by utilising a PTC and a 0.37 kW steam motor (Pytlinski, 1978). A solitary cylinder streak kettle encased in a twofold walled emptied glass sleeve to diminish heat loss was introduced along the central hub. The framework was intended to raise full steam pressure and delivering steam at 374 °C (Pytlinski, 1978). Later in 1938, he utilised a comparable kettle in Florida to control a 0.15 kW steam motor to get a proficiency of 15.5% and an effectiveness of 11.7%.

2.2 Commercial Collectors

The enthusiasm for sun-oriented focusing innovation was irrelevant for nearly 60 years. In any case, in response to the oil emergency of the seventies, universal consideration was attracted to elective energy sources to enhance non-renewable energy sources, and the improvement of various explanatory trough frameworks was supported.

In the mid-seventies, the U.S. government's Sandia national laboratories and Honeywell International Inc. planned the initial two collectors in the United States. The two collectors were very comparable and were set up to work at temperatures underneath 250 °C. A third American organisation, Westinghouse, had engaged with advancement of the collectors at its Production Technology Centre and adjusted Sandia's structure. In 1975, three troughs were constructed and tried at Sandia. These 3.66 m long and had a 2.13 m wide opening, and a 4 cm distance across glass encased dark chrome covered carbon steel absorber, with a 1 cm cleared annulus. One of them was made of pressed wood shell, and another of fiberglass (Shaner & Duff, 1979). During the eighties, this innovation figured out how to enter the market, and some American organisations such as Acurex Solar Corp., Suntec Systems Corp., Solar Kinetics Corp., General Electric Co., have showcased their PTC models (Kutscher et al., 1982).

In the nineties, the following main characteristics of these collectors are highlighted below along with the characteristics of a PTC that was developed by an Israeli company.

- Acurex Corp. consolidated a glass-silvered reflector with thickness of 0.8 mm and a dark chrome-covered steel tube inside a non-emptied borosilicate glass external cylinder with anti-reflective covering (Kalt et al., 1981; Kesselring & Selvage, 1986; Dudley & Workhoven, 1981; Cameron & Dudley, 1987).
- The Acurex Corp. planned and showcased collectors with a shadow-band sun-based following sensor with a multi-component comprising of four photograph identifiers, a focal shadow-band lined up with the central line and a glass window with high-transmittance walled in area shielding the unit from the surrounding (Kalt et al., 1981; Carlton, 1981). Figure 2.1 shows the front and rear views of the Acurex 3001 collector introduced at the Plataforma Solar de Almerı'a (PSA) (Spain).
- The Solar Kinetics collector models (T-700 and T-800) were constructed from aluminised acrylic film and a dark chrome-covered absorber pipe inside a non-emptied borosilicate glass pipe. The T-700 collector was made of aluminium monocoque and second-surface silvered reflector and it has 41.3 mm distance across absorber tube with optical productivity of 0.776. The T-800 collector was made of was spot-welded steel sheet monocoque and 31.8 mm diameter of the absorber pipe with optical productivity of 0.732 (Dudley & Workhoven, 1981; Cameron et al., 1986).
- The Suntec Systems Inc. model collector was a dark chrome-covered steel absorber pipe encompassed by a glass external cylinder containing argon gas vacuum and its reflector was made of surface silvered glass reflect having a thickness of 4.8 mm upheld with copper and Kraton. The structure was supported by a steel pipe pine having a 20.3 cm diameter and steel-sheet ribs (Cameron et al., 1986).

(a)   (b)

Fig. 2.1   The Acurex 3001 collector, (**a**) Front view, (**b**) Rear view. (Kalt et al., 1981; Carlton, 1981)

- The Solel Solar Systems (Israel) collector model was made of hardened steel encompassed by a non-emptied glass envelope with an anti-reflective covering and it has additionally a programmed cleaning framework (Rotemi, 2009) and (Solel, 2009).

These PTCs were at first produced for IPH applications, notwithstanding, the makers discovered three obstructions to fruitful promoting of their innovation. Firstly, a moderate solid promoting and designing exertion was required. Secondly, most potential mechanical clients had unwieldy basic leadership forms, which brought a negative choice after impressive efforts had been consumed. Thirdly, the pace of IPH ventures revenue did not constantly meet the industry criteria (Price et al., 2002).

At that point, Europe had additionally started to build up the PTC innovation with more efforts than in the United States. Maschinenfabrik Augsburg-Nümberg (M.A.N.), in Munich (Germany) was the pioneer organisation creating and showcasing PTCs. They have developed two PTC models, M-480 (one-hub tracking) and Helioman 3/32 (two-pivot tracking) as appeared in Fig. 2.2.

The Helioman 3/32 is the first two-pivot tracking collector and has higher proficiency. However, it has a few significant drawbacks in their mechanical multifaceted nature and thus higher upkeep costs, more failures and less working time with high wind loads, and higher thermal losses.

Many PTC systems are found in the open literature but the most important are those used in CSP plants around the world. The main features of the most widely utilised PTC systems are displayed in Table 2.1.

Fig. 2.2 Helioman 3/32 collector. (Fernandez-García et al., 2010)

Table 2.1 Parabolic trough models used in CSP plants

Developer	Model	Year	Structural design	Aperture Length (m)	Length (m)	Concentration ratio	Optical efficiency	Mirror reflectance
Luz International	LS-2	1985	Torque tube	5	47	71	0.737	0.94
	LS-3	1989	V-truss	5.76	99	82	0.8	0.94
Eurotrough Consortium	ET100	2002	Torque box	5.76	100	82	0.8	0.94
	ET150	2002	Torque box	5.76	150	82	0.8	0.94
ENEA	ENEA	2004	Torque tube	5.76	100	75–80	0.78	NA
Acciona	SGX-2	2005	Aluminum struts	5.76	100–150	82	0.77	NA
SENER	Sener trough	2005	Torque tube	5.76	150	80	NA	NA
Abengoa solar	Astr0	2007	Torque box	5.76	150	NA	NA	NA
	Phoenix	2009	Aluminum struts	5.76	150	NA	NA	NA
	E2	2011	Steel struts	5.76	125	82	NA	NA
TSK Flagsol	Skal-ET	NA	Torque box	5.77	148.5	82	0.8	NA
	Helio trough	2009	Torque tube	6.77	191	76	NA	NA
	Ultimate trough	NA	Torque box	7.51	246	NA	0.89	0.94
IST Solucar	PT-2	2010	NA	4.4	148.5	63	0.75	NA
SkyFuel Inc.	Sky trough	2010	Aluminum struts	5.7	115	72	0.76	0.94
Urssa Energy Co.	Urssa trough	2011	Torque tube	5.76	150	82	0.768	0.93
Albiasa	AT 150	NA	NA	5.77	150	NA	NA	NA
Solarlite	SL 4600	NA	NA	4.6	12	66	0.75	NA
Gossamer Space Frames/3 M	LAT 73	2012	Struts	7.3	192	103	NA	≈0.95

NA No data available
Lupfert et al. (2001, 2003), Günther et al. (2011), Fernández-García et al. (2010), Valenzuela et al. (2014), Channiwala and Ekbote (2015), Rifflemann et al. (2014)

2.3 Background of PTC

The creation of PTC frameworks goes back to the last quarter of nineteenth century. The primary frameworks were utilised in small size offices with lower than 100 kW, similar to steam production and water system. The PTC innovation was popularised in the late seventies and was sent for commercialisation during the eighties (Fernandez-Garcia et al., 2010). In later years, a few organisations fabricated and promoted various PTCs for modern power generation and heat applications. For the period of 1983–1992, nine sunlight-based energy frameworks having 5–85 MWe in size and limit of 354 MWe, were created in Mojave Desert (Price et al., 2002). These frameworks have been active, and their operational encounters have added to PTCs business and developments.

Notwithstanding, the normal yearly development pace of the PTC establishments was right around zero from 1999 to 2006 because of various obstructions against the dispersion of the innovation. The development of the CSP plants rose again in 2006 with 12 MW plant in Spain and 65 MW plant in Nevada. In year 2007 alone, around 90 frameworks for modern heat application systems were accounted for in 21 nations with limit of 25 MWth. Before the finish of 2014, the quantity of introduced modern plants arrived at 124 everywhere throughout the world with a total limit more than 93 MWth (Yilmaz & Soylemez, 2016). There are at present several megawatts under development and a large number of megawatts a work in progress around the world. Several countries such as Algeria, Egypt and Morocco have manufactured coordinated sunlight-based plants. While other countries are finishing or arranging their final plants such as Australia, China, India, Iran, Israel, Italy, Mexico, South Africa and United Arab Emirates (International Energy Agency, 2009). Nowadays, there are in excess of 97 plants at various degrees of improvement dependent on the allegorical trough innovation as indicated by NREL database (International Energy Agency, 2009).

2.4 Basics of PTC

A PTC is a line-centre concentrator, which can be used to change sun-powered energy into high-temperature heat to a temperature up to 550 °C (Tian & Zhao, 2013). As shown in Fig. 2.3, the PTC system fundamentally has a few subsystems to be practically worked. A sun-based illustrative trough framework comprises of an explanatory formed mirror or reflector bended looking like a parabola which in this manner permits focusing the sun's beams (sun-powered radiation) onto the central line of direct beneficiary framework. The mirror is created from various materials such as glass or aluminium to decrease the assimilation losses. The reflectivity of the mirror, its cost, sturdiness and abradable properties are significant variables in the mirror's creation process. A lot of assembling procedures needs to be conducted

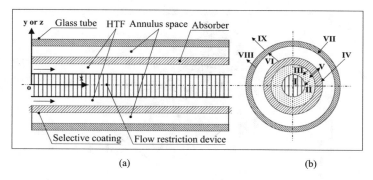

Fig. 2.3 (**a**) Schematic diagrams of (**a**) tube receiver's heat transfer mechanisms, (**b**) PTC's photo-thermal computational model. (Price et al., 2002)

in the case of twisting the mirror such as silvering, defensive covering and sticking to improve the mirror's reflectivity (Behar et al., 2015).

All beams parallel to the collector's central plane are reflected onto the central hub of the collector. It is fundamentally made of hardened steel tube absorber and an envelope made of borosilicate glass encompassing the absorber. The sunlight-based heat moves to a working liquid circulating through the absorber. The absorber tube is coated with extraordinarily coatings to have a high absorbance of sun-based radiation and low emittance of infrared radiation. A glass spread cylinder is concentrically set around the recipient cylinder and emptied to limit the convective and radiative heat losses from the collector. The envelope is covered by hostile to intelligent layer to diminish the heat losses by infrared radiation. Additionally, the space between the absorber and glass cylinders is considered emptied to a very low vacuum pressure of around 0.013 Pa as appeared in Fig. 2.3 (Price et al., 2002).

Getters are usually utilised to guarantee that no hydrogen particles penetrate into the vacuum annulus. The PTC's mirror should be kept stable by bolstering the system by outline with arches with supports to hold the heating element. A control unit is used to drive the collector system by a complete driving setup (apparatus, jackscrew or pressure driven actuator) to situate the collector correctly. The temperature at the central line of a PTC can be as high as 400 °C with a focus proportion of up to 40. The PTC can be orientated either in an east-west bearing, following the sun from north to south, or a north-south heading, following the sun from east to west (Yilmaz, 2018). ASHRAE standard 93 is used as a test method to assess the PTC's thermal performance, which can be applied under indoor or outdoor conditions.

Geometrical relations could be utilised to acquire the geometry of the collector by characterising the explanatory shape that makes up the collector framework (Duffie & Beckman, 1991). The PTC's profile is characterised as shown in Fig. 2.4:

$$x^2 = 4\,yf \qquad\qquad (2.1)$$

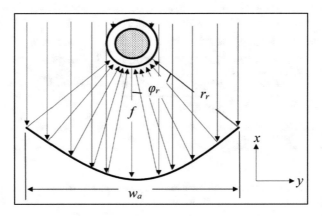

Fig. 2.4 Linear PTC-cross section. (Duffie & Beckman, 1991)

Where f is the focal length, which estimates the position of the heating element, can be calculated by:

$$f = \frac{w_a}{4} \tan\left(\frac{\varphi_r}{2}\right) \tag{2.2}$$

Where w_a is the PTC' aperture width and φ_r its rim angle and it can be calculated by:

$$\varphi_r = \tan^{-1}\left[\frac{8(f/w_a)}{16\left(\dfrac{f}{w_a}\right)^2 - 1} \right] = \sin^{-1}\left(\frac{w_a}{2r_r}\right) \tag{2.3}$$

Where r_r is the rim radius. The radius of the local mirror is calculated by:

$$r = \frac{2f}{1 + \cos\varphi} \tag{2.4}$$

This provides the rim radius r_r as $\varphi = \varphi_r$:

$$r_r = \frac{2f}{1 + \cos\varphi_r} \tag{2.5}$$

The heating element diameter is expected to block all the sun-oriented and it is then calculated by:

$$D = 2r_r \sin 0.267 = \frac{w_a \sin 0.267}{\sin \varphi_r} \tag{2.6}$$

Where 0.267° indicates the half edge of the cone of light emission radiation. The geometrical relations displayed in Eqs. (2.1, 2.2, 2.3, 2.4, 2.5, and 2.6) are very useful in the plan and development of PTC frameworks, which provides precise results. For the examination of PTC frameworks, particularly the optical investigation utilising beam following strategies, is additionally foremost.

There are significant definitions for assessing the thermal, and optical performances of PTCs, which are exhibited and discussed below. Both qualitative and quantitative reviews of energy can be completed by investigating the energy and exergy of PTC frameworks.

2.4.1   Optical Analysis

The optical proficiency, η_o is characterised as the proportion of the energy consumed by the heating element to the energy on the collector's opening and it is defined by:

$$\eta_o (\theta = 0) = \rho \tau \alpha \gamma \tag{2.7}$$

Where (ρ) is the capacity of the reflectivity of the mirror, (τ) is the transmittance of the glass envelope, (α) is the absorptivity of the covering on the absorber surface and (γ) is the capture factor of the mirror and heat element.

The variety of all optical properties rely upon on the impact of the rate edge of sun-oriented collectors which is related to a modifier called the frequency edge modifier and it can be calculated by (Gaul & Rabl, 1980),

$$K(\theta) = \frac{\eta_o (\theta)}{\eta_o (\theta = 0)} \tag{2.8}$$

The incidence angle varies as per the following mode applied. It is important that the episode point can be depicted by different connections given by (Yilmaz & Mwesigye, 2018) contingent upon the tracking kind. The optical effectiveness incorporates the impact of incidence angle and the end-loss factor, is displayed as:

$$\eta_r (\theta) = \rho \tau \alpha \gamma \Gamma \cos \theta \tag{2.9}$$

The end-loss factor, Γ is estimated using (Gaul & Rabl, 1980):

$$\Gamma = 1 - \frac{f}{l} \left(1 + \frac{w_a^2}{48 f^2} \right) \tan \theta \tag{2.10}$$

Equation (2.10) is appropriate for equivalent lengths of the heating element and the collector when heat element is considered evenly. However, the end-loss factor is calculated when the heating element length stretches out past the collector length l by a sum r on one side, using:

$$\Gamma = 1 + \frac{r}{l} - \frac{f}{l}\left(1 + \frac{w_a^2}{48f^2}\right)\tan\theta \qquad (2.11)$$

The end-loss impact for an on a level plane situated north–south pivot framework is resolved in detail by (Xu et al., 2014). The end-loss for short trough collectors was proposed by (Xu et al., 2014) using different technique. The end-loss in tube shaped troughs is given by (Edenburn, 1976) using an alternate way.

The optical plan of the trough collector is practically influenced by a few elements (Guven & Bannerot, 1986) including: changes in the sun's width and occurrence angle impacts, thermophysical properties of the materials utilised in heating element and mirror development, blemishes (or blunders), flawed following of the sun and poor working strategies. (Mokheimer et al., 2014) have envisioned the impacts of the PTC's parts on the optical effectiveness as appeared in Fig. 2.5. Taking note of that distinguishing the incomplete impacts of these elements will explain the assurance of the optical productivity.

During the optical examination of PTC frameworks, various factors and their impacts need to be explored such as the assurance of the catch factor, optical

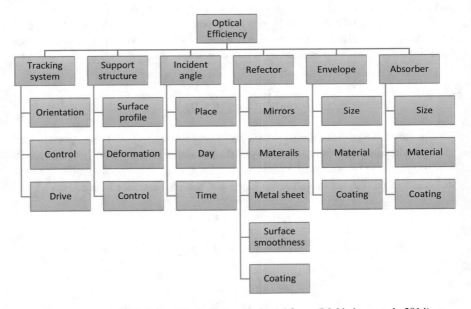

Fig. 2.5 Parameters affecting the optical efficiency adapted from. (Mokheimer et al., 2014)

productivity and the heat flux departure from the heating element, rim angle, heating element size, optical error, sun shape, etc.

The thermal examination of PTC from the energetic and exergetic points of view as well as the entropy generation are discussed next.

2.4.2 Thermal Performance

The PTC's thermal efficiency is calculated by:

$$\eta_{th} = \frac{Q_u}{A_{ap} I_D} \tag{2.12}$$

Where Q_u is the useful energy transferred to the working fluid:

$$Q_u = \dot{m} C_{p,f} \left(T_{out} - T_{in} \right) \tag{2.13}$$

The thermal enhancement factor (η) evaluates the effectiveness of heat transfer and flow enhancement methods (Mwesigye et al., 2018).

$$\eta = \frac{\left(Nu / Nu_p \right)}{\left(f / f_p \right)^{1/3}} \tag{2.14}$$

The Nusselt number could be evaluated by:

$$Nu = \frac{h D_{ri}}{k} \tag{2.15}$$

where h is the heat transfer coefficient between working fluid and the absorber tube.

$$h = \frac{Q_u}{\left(\pi . d_{ri} . L \right) \left(T_r - T_{fm} \right)} \tag{2.16}$$

and,

$$T_{fm} = \frac{T_{in} + T_{out}}{2} \tag{2.17}$$

The friction factor could be determined by:

$$f = \frac{\Delta p}{\frac{1}{2}\rho u^2}\left(\frac{d_{ri}}{L}\right) \tag{2.18}$$

2.4.3   Exergetic Performance

The exergetic effectiveness can be assessed using the pressure losses and the thermal contribution to PTC:

The entropy generation is calculated by (Mwesigye et al., 2018):

$$\dot{S}_{gen} = \left(\frac{dS}{dt}\right)_{CV} - \sum_{i=0}^{n}\frac{\dot{Q}_i}{T_i} - \sum_{0}\dot{m}_o s_o \geq 0 \tag{2.19}$$

The exergy rate is evaluated by:

$$Ex_u = \dot{m}C_{p,f}\left[\left(T_{out} - T_{in}\right) - T_{amb}\ln\left(\frac{T_{out}}{T_{in}}\right)\right] \tag{2.20}$$

The solar radiation exergy is given by:

$$Ex_a = A_{ap}I_D\left[1 + \frac{1}{3}\left(\frac{T_{amb}}{T_s}\right)^4 - \frac{4T_{amb}}{3T_s}\right] \tag{2.21}$$

The exergy efficiency is calculated by:

$$\eta_{exr} = \frac{Ex_u}{Ex_a} = \frac{\dot{m}C_{p,f}\left[\left(T_{out} - T_{in}\right) - T_{amb}\ln\left(\frac{T_{out}}{T_{in}}\right)\right]}{A_{ap}I_D\left[1 + \frac{1}{3}\left(\frac{T_{amb}}{T_s}\right)^4 - \frac{4T_{amb}}{3T_s}\right]} \tag{2.22}$$

2.4.4   Entropy Analysis

In any numerical analysis, comprehending the governing equations would give the dispersions for speed, temperature, pressure and turbulent amounts inside the absorber tube (Herwig & Kock, 2007). The entropy generation comes from irreversibilities of heat transfer $(S'''_{gen})_H$ and fluid friction $(S'''_{gen})_F$ and its formula is given:

$$S'''_{gen} = \left(S'''_{gen}\right)_F + \left(S'''_{gen}\right)_H \tag{2.23}$$

The entropy generation caused by irreversibility of fluid friction is calculated by:

$$\left(S'''_{gen}\right)_F = S'''_{PROD,VD} + S'''_{PROD,TD} \tag{2.24}$$

Where $S'''_{PROD,VD}$ is the direct dissipation entropy production and $S'''_{PROD,TD}$ is the indirect (turbulent) dissipation entropy production

$$S'''_{PROD,VD} = \frac{\mu}{T}\left(\frac{\partial u_i}{\partial x_j} + \frac{\partial u_j}{\partial x_i}\right) \tag{2.25}$$

$$S'''_{PROD,TD} = \frac{\rho\varepsilon}{T} \tag{2.26}$$

The entropy generation due to irreversibility of heat transfer is calculated by:

$$\left(S'''_{gen}\right)_H = S'''_{PROD,T} + S'''_{gen,TG} \tag{2.27}$$

Where $S'''_{PROD,T}$ is the heat transfer entropy production using mean temperatures and $S'''_{gen,TG}$ is the heat transfer entropy production using fluctuating temperatures

$$\text{Where } S'''_{PROD,T} = \frac{\lambda}{T^2}(\nabla T)^2 \tag{2.28}$$

$$S'''_{gen,TG} = \frac{\alpha_t}{\alpha}\frac{\lambda}{T^2}(\nabla T)^2 \tag{2.29}$$

Where α and α_t are the thermal diffusivities.

The entropy generation calculation using the direct method (Herwig & Kock, 2007) was given in Eqs. (2.23, 2.24, 2.25, 2.26, 2.27, 2.28, and 2.29). The entropy generation is calculated using with the indirect method given by (Bejan, 1996) as:

$$S'_{gen} = \frac{\overset{\cdot}{q}^2}{\pi\,\lambda\,T_{bulk}^2\,Nu} + \frac{32\,\overset{\cdot}{m}^3\,C_f}{\pi^2\,\rho^2\,T_{bulk}\,D^5} \tag{2.30}$$

where q is the heat transfer rate per unit length, $Nu = h\,D/\lambda$ with $h = q/\left(T_w - T_{\text{bulk}}\right)$, $C_f = \left(-\dfrac{dp}{dx}\right)\rho D / 2G^2$, with $G{=}4\ m/\pi D^2$ and T_{bulk} is the bulk fluid temperature $T_{\text{bulk}} = (T_{\text{out}} + T_{\text{in}})/2$.

The entropy generation rate for a fluid occupying a volume V is calculated by:

$$S_{gen} = \iiint_V S'''_{gen}\, dV \tag{2.31}$$

The Bejan number, presents the irreversibility caused by heat transfer to total entropy generation, is described as:

$$Be = \frac{\left(S'''_{gen}\right)_H}{S'''_{gen}} \tag{2.32}$$

The irreversibility of heat transfer is predominant when $Be = 1$, and the irreversibility of fluid friction is prevailing when $Be = 0$ (Amani & Nobari, 2011).

2.5 PTC Applications

The PTC systems are utilised in many industrial applications, and it is classified into three main groups (heating, cooling and chemical/physical) as shown in Fig. 2.6. The heat collected is used directly for a given process in thermal applications. The solar energy is primarily utilised in heating and cooling applications in the industrial and commercial sectors, but concentrating PVs are currently under development. Each of these application groups is described in the following sections.

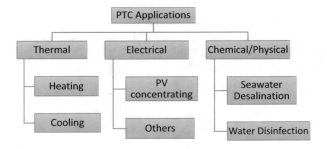

Fig. 2.6 Classification of PTC applications

2.5.1 *Thermal Heating*

PTC systems are utilised in a variety of industrial applications. The application of solar energy in the heating sector has received an intensive attention by many researchers. The collector transfers heat to the HTF, which is used as a source of energy for a given process (heating a fluid as the main objective of the PTC system). Heating applications can be classified into two groups based upon the temperature reached by the HTF: (i) Low-temperature applications are for a maximum temperature of 100 °C and (ii) medium-temperature applications usually reach temperatures up to 400 °C. Low-temperature PTC systems are usually utilised in preheating and drying processes in commercial, residential and industrial sectors. Steam generation (SG) and concentrating solar power (CSP) are the principal applications for medium-temperature PTC systems. (Kalogirou, 2004) provided a list of potential industrial processes for deploying PTC systems, and Table 2.2 shows some of the processes. The temperature ranges of the processes are in the operational range of temperature of the PTC systems.

CSP is the most popular PTC application for electricity generation. CSP plants generate electricity using solar-assisted turbines or integrated systems with combined cycles, usually with a thermal storage block. Both direct and indirect SGs are the most commonly used methods in the world for electricity generation with PTCs. These systems generate steam to run a turbine, by direct heating (water as HTF of the PTC system) or by indirect heating (i.e., oil as HTF and heat is transferred to water through a heat exchanger) as shown in Fig. 2.7. The thermal storage block increases the efficiency and the capacity factor, making it possible to generate electricity even when there is no solar resource (nights or cloudy days). Cogeneration plants are another potential application for heating PTC systems and its energy costs and CO_2 emissions are lower than conventional cogeneration plants.

2.5.2 *Thermal Cooling*

Solar cooling refers to the use of solar radiation as a thermal energy source to cool a fluid or a space. Absorption and adsorption are two main solar cooling methods and both methods will replace the mechanical compressor in a conventional cooling process with a thermal compressor. The absorption process is the most commonly used method for solar cooling with PTCs. Absorption uses a fluid-fluid mixture (also called working pair) as the refrigerant. The fluids of the working pair make a strong solution when mixed at low temperatures and can be separated when the mixture is heated. The solute is converted into a gas and the solvent remains in liquid state when the mixture is heated. The absorption cycle takes place as follows:

(i) The mixture is separated in the generator by heating.

Table 2.2 Temperature range for industrial processes (Kalogirou, 2004)

Industry	Process	Temperature (°C)
Dairy	Pressurisation	60–80
	Sterilisation	100–120
	Drying	12–180
	Boiler feed water	60–90
Tinned food	Sterilisation	110–120
	Pasteurisation	60–80
	Cooking	60–90
	Bleaching	60–90
Bricks and blocks	Curing	60–140
Plastics	Preparation	120–140
	Distillation	140–150
	Separation	200–220
	Extension	140–160
	Drying	180–200
	Blending	120–140
Chemical	Soaps	200–260
	Synthetic rubber	150–200
	Processing heat	120–180
	Preheating water	60–90
Paper	Cooking, drying	60–80
	Boiler feed water	60–90
	Bleaching	130–150
Heating of buildings	–	25–75
Desalination	–	100–250
Power cycles	Vapor generation	300–400
	Phase-change materials	300–375
Others	General steam generation	130–210

(ii) The solute (gas) is condensed, and it rejects heat to the ambient space. The solute is then expanded and later evaporated by heat collected from the refrigerated space (as in traditional cooling systems).

(iii) The heated solute is then mixed with the solvent in the absorber.

(iv) The mixture is pumped to the generator and preheated by a heat exchanger with pure solvent coming from the generator.

(v) The mixture is heated again in the generator, closing the cycle.

The most common processes in absorption cooling are the single-effect and double-effect cycles. Figure 2.8 shows a schematic diagram of single and double-effect absorption cycles. The most commonly used working mixture in solar absorption systems are lithium bromide and water ammonia.

Solar adsorption cooling processes are entirely different from the absorption processes. The physical principle behind adsorption cooling is a surface-based phenomenon in which a porous material (adsorbent) captures vapor from a fluid (refrigerant) and the adsorbent is regenerated by heating. Adsorption processes

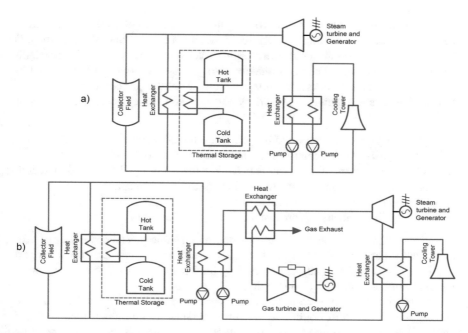

Fig. 2.7 (**a**) Direct SG CSP plant diagram, (**b**) Indirect SG CSP plant diagram integrated with a combined cycle. (Kalogirou, 2004)

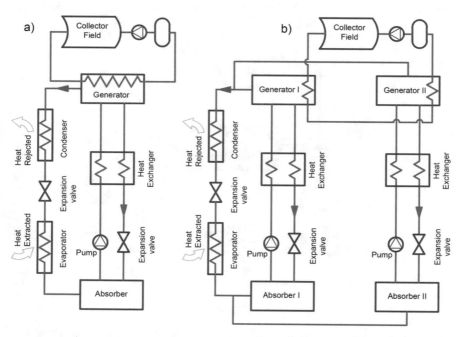

Fig. 2.8 Solar-assisted absorption diagrams: (**a**) single effect, (**b**) double effect. (Tagle-Salazar et al., 2020)

differ from absorption because the working mixture consists of a solid–fluid combination and moreover heating is intermittent, not a continuous one (the adsorbent is heated whenever it is saturated). Adsorption cycles require very low or no mechanical or electrical input but need thermal input (e.g., from the collector field) which is very important, and it works intermittently with the solar resource. A simple adsorption system cycle diagram is shown by Fig. 2.9 and the cycle of working is as follows:

(i)   The refrigerant is evaporated by heat from the refrigerated space.
(ii) The vaporised refrigerant is adsorbed in the adsorption chamber.
(iii) When the adsorption chamber gets heated, the vapor is released and then condensed, resulting in the rejection of heat to the ambient space.
(iv) The condensates are stored in a tank.
(v)   Finally, the condensates are evaporated again, closing the cycle.

In solar adsorption cooling, nonconcentrating technologies (such as flat plates or evacuated tubes) are most frequently used. Nevertheless, some literatures suggest good performance for an adsorption cooling pilot plant with PTC technology compared to nonconcentrating technologies.

The most commonly used adsorbent materials are zeolite, activated carbon and silica gel, and the most commonly used refrigerants are ammonia, methanol and water. Chemical adsorption is another method for cooling. It is based on a reaction between the adsorbent and refrigerant that form a strong chemical bond, increasing heat transfer. The main disadvantage of chemical adsorption is that it cannot be easily reversed as higher amount of energy is needed to break the chemical bond, and

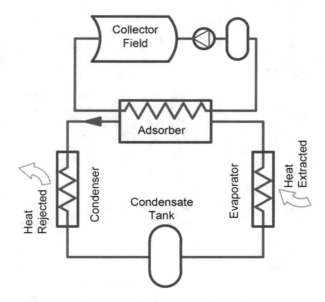

Fig. 2.9 Simple intermittent solar-assisted adsorption diagram. (Tagle-Salazar et al., 2020)

the chemical reaction alters the state of both adsorbent and refrigerant in a continuous operation. The most used adsorbent for chemical adsorption is calcium chloride, which can make a working pair with either water or ammonia.

The general advantage of solar cooling technologies is lower energy consumption compared to the conventional vapor-compressor systems. The working pair (in a liquid phase) is pumped in solar absorption cooling rather than using a compressor as in conventional cooling processes,

and there is almost no mechanical input in adsorption cooling, as described earlier. Thus, the energy consumption is drastically reduced. Other advantages are low noise and low vibration because of fewer moving parts. The principal disadvantage is their low COP, with a COP of around 0.7 for single-effect absorption, 1.2 for double-effect absorption and 0.1–0.2 for adsorption compared with a COP of 3–4 for conventional vapor-compression systems (Sarbu & Sebarchievici, 2013).

2.5.3 Seawater Desalination

The process in which minerals are separated from seawater to produce fresh water is called Seawater desalination. Various methods are utilised to separate salts from seawater in industry, and they are divided into four groups: phase-change processes, single-phase processes, electric processes and hybrid processes as shown in Fig. 2.10. Phase-change processes use thermal energy to separate the brine and the most commonly used methods are multistage flash and multieffect distillation (MED). Single-phase processes use mechanical separation by passing seawater through filter membranes where salts are trapped, and reverse osmosis (RO) and

Fig. 2.10 Seawater desalination processes classification

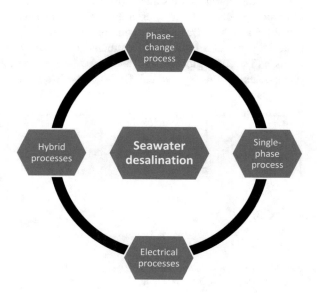

forward osmosis are the most frequently used methods. Electric processes are based on cationic and anionic ion-exchange effects. In this electrolysis process, cathode and anode membranes are arranged alternately and exposed to an electric field, so that salt particles are trapped during the process and separated from seawater. The principal methods of electric processes are electro-dialysis, ion exchange and capacitive deionisation. Hybrid processes usually mix phase change with single-phase processes, such as membrane distillation method.

Solar seawater desalination with PTC technology participates directly only in phase-change processes as PTCs provide thermal energy. Brine and fresh water are separated by heating seawater in thermal processes and the process is represented in Fig. 2.11a. Seawater is heated so that water is vaporised, which is physically separated from salt because of its lower boiling point. The water vapor then condenses, leaving fresh water. A Rankine cycle provides the necessary electrical energy to pressurise the seawater and pump it through the membranes in single-phase processes, as shown in Fig. 2.11b. The Rankine cycle operates as a CSP plant. Both thermal processes and membrane processes have advantages and disadvantages over each other. Thermal systems' advantages over membrane systems: proven and established technology, higher quality product, less rigid monitoring required than for membrane processes, less impacted by quality changes in feed water and no membrane replacement costs.

2.5.4   *Water Decontamination*

The process of removing hazardous compounds such as heavy metals, organic compounds and chemical substances from water is called Decontamination. Advanced oxidation processes (AOPs) offer a feasible and sustainable alternative for disinfection of water. General applications of AOPs include the degradation of resistant material prior to a biological treatment and treatment of refractory organic compounds (Tchobanoglous et al., 2003). AOPs generate a high concentration of oxidants (usually hydroxyl radical, $OH-$) to oxidise polluting matter that would not be

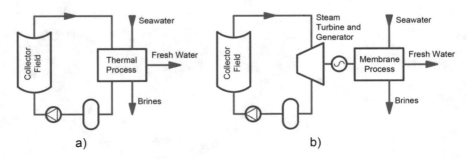

Fig. 2.11 (**a**) phase-change process for seawater desalination, (**b**) single-phase process for seawater desalination. (Tagle-Salazar et al., 2020)

easy to separate by biological degradation. There are some AOPs that use UV radiation as an energy source to produce the oxidants. But heterogeneous photocatalysis (HPC) with TiO_2 and photo-Fenton process (PFP) is the most commonly used methods in solar applications. These processes generate hydroxyl radicals (*OH) when UV radiation activates the catalyst in an atmosphere with oxygen.

Solar decontamination is usually realised in batch mode as shown in Fig. 2.12. General purposes of water decontamination are performed for drinking water or agricultural applications. Initially the catalyst is mixed in water to suspend particles in the fluid and then the mixture is pumped to the solar field to undergo the chemical processes before returning to the mixer. This process repeats continuously until the pollutants are degraded. The solar photocatalysis is a promising technology when compared to others because of its low impact.

Table 2.3 compares various technologies used for water decontamination. It is not very difficult to compare solar decontamination systems against conventional systems, which is an obstacle to industrial application.

2.5.5   Concentrating PVs

Concentrating PV (CPV) generation is the direct conversion of solar energy into electrical energy using semiconductor materials. Photoelectric effect is the basic principle behind the operation of the solar cell and that effect generates a potential difference within a semiconductor when it is exposed to sunlight. PV concentrator (PVC) systems with PTCs have the same equipment as flat-plate PV collectors such as converters, batteries and voltage regulators because both collectors provide direct current.

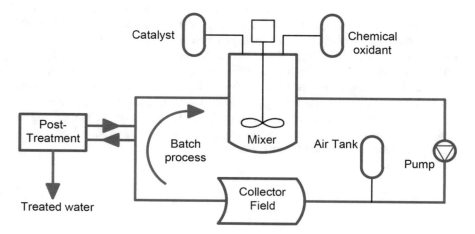

Fig. 2.12 Schematic diagram of solar-assisted wastewater treatment plant. (Blanco et al., 2009; Malato et al., 2004)

Table 2.3 Comparison of collector technologies for decontamination processes

Technology	Advantages	Disadvantages
PTC	Smaller size of reactors High volumetric flow (turbulent flow) Better mass transfer Low catalyst load No vaporisation of compounds	It needs direct solar beam radiation High costs Mayor optical losses Low water overheating Low quantum efficiency
NCC	High optical efficiency Use direct and Diffuse radiation Design simple and easy Low cost No overheating	High sizes of reactors Low mass transfer and volumetric flow (only laminar flow) Reactant evaporation and contamination (if open) Weather resistance, chemical inertness and UV transmission
CPC	Smaller size of reactors High volumetric flow (turbulent flow) Better mass transfer Use direct and diffuse radiation Low catalyst load	Difficult to scaled up Moderate heat generation Moderate capital cost

Table 2.4 Characteristics of concentrating photovoltaic applications

Concentration type	Concentration factor	Collector type	Cell type
High	> 400	PD	Multijunction
Medium	3–100	PTC, LFC	Silicon and others
Low	3	CPC	Silicon

PVC operates in the same way as thermal concentrator but works with a modified receiver with PV cells on its surface. Concentrated sunlight strikes the PV cells, and PV cells convert solar radiation into electricity. The cells generate more energy under concentrated sunlight due to the incident energy density. It is a fact that PV cells do not convert all the incident energy into electricity and most of this rejected energy is converted into heat, which lead to temperature of the cell to increase, affecting its efficiency. The fluid used in the cells will be flowing inside the receiver. These collectors are known as thermal-PVCs (T-PVCs). PV cells used in concentrating PVs are designed to resist high incident radiation, so the use of Si-based cells fully depends on the concentration factor of the PVCs, as shown in Table 2.4. The main advantage of PVCs over nonconcentrating PV systems is that they have a higher efficiency and require fewer PV cells.

Chapter 3
PTC Enhancement Using Passive Techniques

3.1 Introduction

Effective heat transfer framework is one of the significant prerequisites in energy protection (Hojjat et al., 2011). The improvement of heat transfer prompts growth of high heat motion. Aside from this, improvement in heat movement rate likewise prompts a few focal points such as decrease of heat exchanger size and temperature main thrust, etc. The decreased size of heat exchanger is very significant from financial perspective, although decrease of temperature main thrust prompts increment second law proficiency and minimisation of entropy generation or least energy annihilation. The high heat transfer is also useful because of the way that heat exchanger can be worked at low speed and gives impressively higher heat transfer coefficient. Therefore, low working pressure drop is accomplished and working expense is extensively diminished. Thus, to improve the productivity of heat exchangers, it is imperative to improve the thermal contact and reduce the pumping power. These advantages related with heat transfer improvement powers to investigate various systems/techniques to improve heat execution of heat exchangers.

3.2 Passive Techniques

The idea of thermal improvement is very significant and valuable in control, refrigeration, cooling, car enterprises, etc. Furthermore, heat transfer improvement systems are likewise turning into a significant matter of enthusiasm for electronic cooling, sunlight-based heat collectors, small scale synthetic preparing smaller heat exchanger structure and so forth (Geyer et al., 2007).

The issue of heat transfer improvement has gotten increasingly indispensable in every single mechanical application. The heat transfer upgrade systems can be

Fig. 3.1 Heat transfer
augmentation strategies

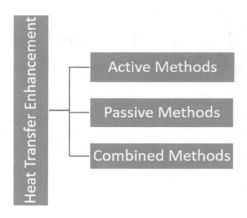

classified into three categories: active, passive and compound techniques as shown in Fig. 3.1. In active strategy, external power is utilised for heat transfer upgrade. It appears to be a simple strategy in a few applications, but it is very perplexing from configuration perspective. That is the reason it is of restricted use because of external power necessities. Passive strategies use energy inside the framework which prompts increment liquid pressure drop (Dewan et al., 2004). The utilisation of uncommon surface geometry gives high heat execution when compared with smooth surface. Surface modifications or alterations such as twisted tapes, inserts, coils, ribs, dimples, blades, fins and so forth are distinctive passive types, which are utilised to upgrade heat transfer rate. Likewise, tube with longitudinal supplements is additionally a viable passive technique for heat transfer improvement (Hsieh & Huang, 2000). Passive techniques are commonly metallic strips, which are turned in some particular shapes/measurements and embedded over the flow stream. They are additionally deemed as swirl stream gadgets as turbulators used to confer the swirl stream, which prompt the expansion in heat transfer coefficient. Prior, it was hard to work with complex geometries because of their creation requirements; however, with the headway in assembling innovation, it is presently very conceivable to apply new geometries in heat transfer upgrade strategies. Compound heat transfer strategy is a half-breed system, which includes the utilisation of both active and passive techniques, which is very unpredictable with restricted applications.

Passive strategy is obtained by creating swirl stream of the cylinder side liquid, which gives high speeds close to limit and liquid blending and thus high heat transfer coefficient. In heat transfer frameworks outfitted with aloof procedures, the heat transfer and pressure drop attributes are represented by passive strategy geometrical factors. Additionally, little clearances between a passive system and cylinder limit are significant factor while choosing the width of the passive strategy, which can create sidestep stream, which lead to pressure drop. Utilisation of passive procedure, for instance twisted tapes and different supplements, causes stream blockage, stream portioning and enlistment of auxiliary stream. Free stream territory is decreased because of stream blockage, pressure drop and viscous impacts are extensively diminished. Furthermore, flow speed additionally increases and, in many cases, secondary stream is actuated. This secondary stream generates swirl and gives compelling blending of liquid stream which enhances the temperature gradient and along these lines heat transfer coefficient (Dewan et al., 2004).

3.2.1 PTC Studies Using Passive Techniques

Extensive experimental and numerical research has been conducted on thermal enhancement using various strategies, which is thoroughly investigated and summarised in the next sections and listed in Table 3.1.

Brooks et al. (2006) carried out a basic study following ASHRAE 93–1986 standard to investigate the PTC thermal effectiveness. Senthil Manikandan et al. (2012) conducted a theoretical study on a solar PTC thermal performance by varying few factors such as flow velocity, and concentration ratio to estimate the useful exergy and heat removal factor. Lobon and Valenzuela (2013) used sensitivity analysis to examine the thermohydraulic small PTC effectiveness utilising water-steam. The effectiveness was assessed at different inlet temperature pressure, flow velocity and solar incidence. Fernández-García et al. (2015) studied numerically with the aid of CFD tool the thermal performance of various PTC designs utilised for practical process heat applications.

Kalogirou (2012) studied convection heat transfer in different areas of PTC system: in the absorber tube, in the evacuated gap and the glass envelope. It was found that the PTC's efficiency to be 58% at 200 °C operating temperature. Cheng et al. (2015) investigated numerically using 1-dimensional non-uniform model the thermal performance of solar PTC. It was inferred that the nonuniform model delineated minor varieties in the overall thermal effectiveness. The thermal effectiveness of a sunlight-based PTC with theoretical and numerical models were recently investigated by Conrado et al. (2017).

The deformation and nonuniform heat transfer characteristics of PTC absorber was examined by Lu et al. (2016). The PTC's thermal efficiency varied from 57.8% to 65.6%. Additionally, Liu et al. (2010) found experimentally that the PTC's efficiency varied from 40% to 60%. Valan Arasu and Sornakumar (2006) found that the hot water generation system overall efficiency was 48% and the PTC's efficiency

Table 3.1 Layout of metal foams inserted in a receiver tube Wang et al. (2013). (License Number: 4606950012954)

Layout of metal foams inserted in receiver tube.

Type	$h'/D_i=0$	$h'/D_i=0.25$	$h'/D_i=0.5$	$h'/D_i=0.75$	$h'/D_i=1$
Series 1					
Series 2					

was 58%. Kumaresan et al. (2012) studied experimentally the PTC effectiveness combined with thermal energy storage unit. They found that the peak efficiency was 62.5% at 12:00 noon and using Therminol 55. Further study on solar PTC using a control volume analysis with a comprehensive exergetic balance was performed by Padilla et al. (2014). It was observed that the exergetic efficiency increased and the thermal efficiency decreased at high solar radiation. It was pointed out that the flow velocity had an insignificant influence on the overall exergetic efficiency. Moreover, the PTC's exergetic and thermal efficiencies would be contrarily relative under various working conditions (Guo et al., 2016).

The next sections highlight various studies performed chronologically by different investigators on PTC augmentation using different types of passive techniques.

Grald and Kuehn (1989) used innovative cylindrical porous absorber to examine its efficacy for enhancing the PTC's thermal effectiveness. They developed 1D finite difference computer program to compute the fluid velocity and transient temperature distributions along the midpoint of the collector. They determined the influence of mass flow rate, acceptance angle, receiver dimensions and materials properties on the thermal efficiency. Numerical results for thermal efficiency show the potential for marked improvement over available PTCs for a wide range of fluid outlet temperatures. The performance benefits of the proposed design were most pronounced at high fluid outlet temperatures.

The PTC's performance with different porous shapes (square, triangular, trapezoidal and circular) was numerically studied by Reddy and Satyanarayana (2008) as shown in Figs. 3.2, 3.3 and 3.4. The effects of solar radiation, and porous fin geometrical parameters were examined on the PTC's performance. The optimal thermal and fluid flow characteristics were shown using the trapezoidal shape with 4-mm thickness, 0.25 of tip-to-base thickness ratio (λ) and 1 of fin distance to diameter (l/d). The PTC's efficiency remarkably improved with the usage of porous fin receiver and the heat transfer was 13.8% augmented with 1.7 kPa pressure drop losses.

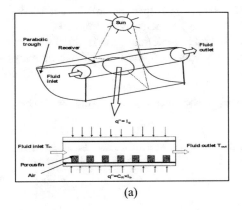
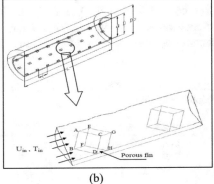

(a)   (b)

Fig. 3.2 (a) Solar PTC with its boundray conditions and (b) Porous fin receiver with interface boundray conditions Reddy and Satyanarayana (2008). (License Number: 4606940724672)

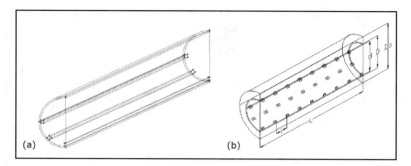

Fig. 3.3 Schematic diagram of Reciever's computational domains: (**a**) longitudinal fin and (**b**) porous fin Reddy and Satyanarayana (2008). (License Number: 4606940724672)

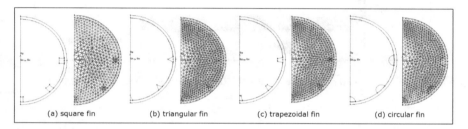

Fig. 3.4 Cross-section of different porous fin receivers with its grid generation Reddy and Satyanarayana (2008). (License Number: 4606940724672)

Reddy et al. (2008) presented a thermal analysis of a porous PTC receiver as shown in Fig. 3.5 considering the heat gain and heat loss caused by free convection. CFD-FLUENT package was used to solve the numerical model with the aid of RNG k-ε turbulent model. Different geometrical fin factors were considered in the thermal model such as thickness, porosity, aspect ratio and heat flux as shown in Figs. 3.6 and 3.7. It was observed that the 17.5% of heat transfer improvement and 2 kPa pressure drop loss were achieved. It is also found that the shorter fins were progressively ideal to improve the thermal and hydraulic characteristics. Porous fin receiver having 4 mm of thickness showed a significant increase thermal performance. Empirical correlation for the Nusselt number is suggested dependent on the numerical results.

A three-dimensional numerical simulation of solar PTC having porous disc receiver was conducted by Kumar and Reddy (2009) as shown in Fig. 3.8. The effects of Therminol-VP1 as a working fluid, geometrical parameters (disc's angle (h), orientation, height (H) and distance (w)) and solar radiation intensity on the overall thermal performance were involved in the thermal analysis as shown in Fig. 3.9. It was found that higher thermal characteristics were obtained using top porous disc receiver with w = di, H = 0.5di and h = 30 at θ = 30°. The numerical outcomes show that 64.3% and 457 Pa were achieved for Nusselt number improvement and pressure drop loss, respectively, compared with the smooth receiver.

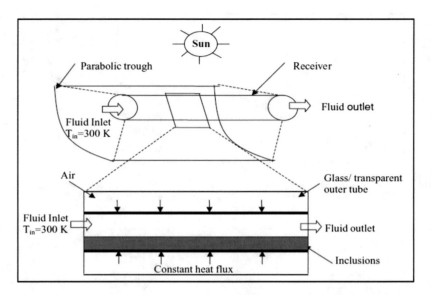

Fig. 3.5 Solar PTC with receiver Reddy et al. (2008). (License Number: 4606940724661)

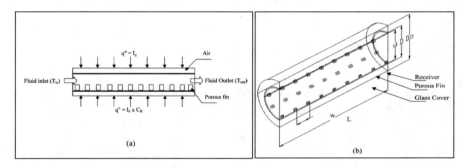

Fig. 3.6 Solar PTC with receiver having porous fin: (**a**) receiver with its operating conditions and (**b**) receiver with its geometrical factors Reddy et al. (2008). (License Number: 4606940724661)

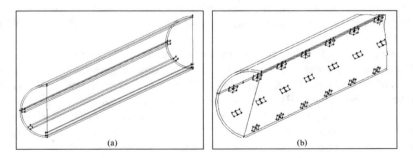

Fig. 3.7 Solar PTC receiver physical model with: (**a**) longitudinal fin and (**b**) porous fin Reddy et al. (2008). (License Number: 4606940724661)

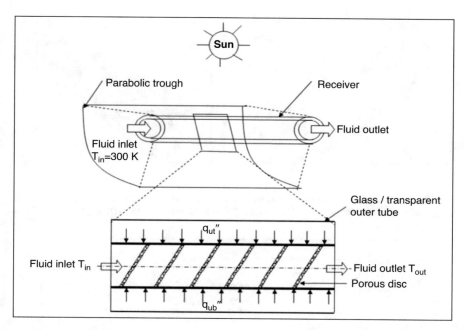

Fig. 3.8 Solar PTC with porous receiver Kumar and Reddy (2009). (License Number: 4606940725336)

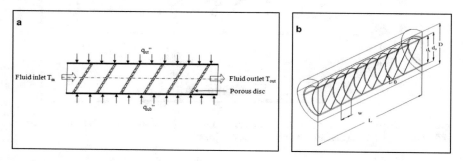

Fig. 3.9 Solar porous receiver: (**a**) with boundray conditions and (**b**) with geometrical parameters Kumar and Reddy (2009). (License Number: 4606940725336)

Munoz and Abanades (2011) analysed numerically using CFD tools the impact of helical finned tubes in PTC configuration as shown in Fig. 3.10. The thermal stress, fatigue, deformation and pressure losses were considered in the analysis. The results of the finned tube configurations were validated with a reference commercial tube. It is shown that the pressure losses in PTC tube increased when the fins number and its helix angle increased. The thermal and exergetic efficiencies of the collector increased with the decrease of thermal losses and temperature gradients as shown in Fig. 3.11.

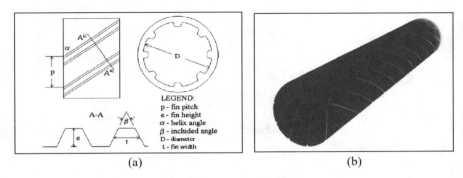

(a)   (b)

Fig. 3.10 (**a**) Geometric variables of the helical fin and (**b**) Tube model corresponding to the length of a helix Munoz and Abanades (2011). (License Number: 4606940725264)

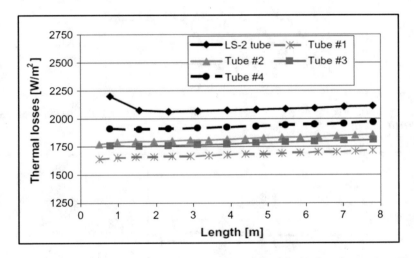

Fig. 3.11 Thermal losses for PTC tube with helical fins compared with the reference case Munoz and Abanades (2011). (License Number: 4606940725264)

Kumar and Reddy (2012) performed a 3D numerical simulation using CFD-FLUENT package for a porous disc enhanced receiver to find the best configuration with water and therminol oil as shown schematically in Fig. 3.12. The performance of the PTC was evaluated by examining the influence of porous disc receiver's location at bottom, top; and alternative porous disc as shown schematically in Figs. 3.13 and 3.14. The porous disc geometric parameters impact and working fluid factors on the PTC's thermal effectiveness were also investigated. It is found that the flow field caused by the solid/porous discs were remarkably affected the local heat transfer coefficient. The results revealed that the low-pressure drop was attained using porous disc PTC receiver compared to solid disc one. The optimum outcomes were attained at 221 W/m enhanced the heat transfer rate with 13.5% pumping losses for porous disc PTC receiver using water. Conversely, there was 575 W/m of heat transfer rate enhancement with 31.4% pumping losses using therminol oil-55 compared

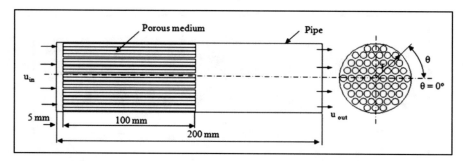

Fig. 3.12 Schematic diagaram of the geometry used Kumar and Reddy (2012). (License Number: 4606940725153)

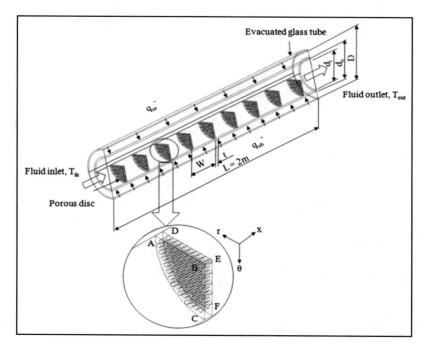

Fig. 3.13 Sectional representation of porous disc PTC receiver Kumar and Reddy (2012). (License Number: 4606940725153)

with smooth PTC receiver. Nusselt number and friction factor empirical correlations were developed for porous disc PTC receiver.

Cheng et al. (2012a, b) carried out a numerical analysis on combined thermal and turbulent flow in a novel PTC absorber tube using unilateral milt-longitudinal vortexes (UMLVG-PTR) as shown schematically in Fig. 3.15. They used finite volume method (FVM) and the Monte Carlo ray-trace (MCRT) method combined with the field synergy principle (FSP). Various parameters such as solar radiation, working fluid inlet temperature, Reynolds number and LVG geometric factors were further studied. It

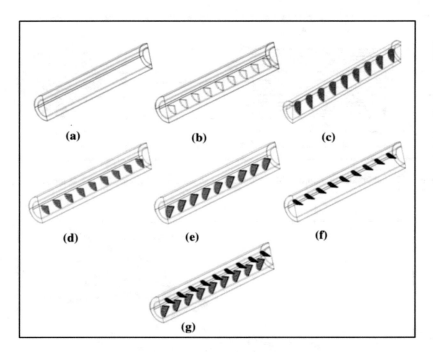

Fig. 3.14 Solar PTC configurations Kumar and Reddy (2012). (License Number: 4606940725153)

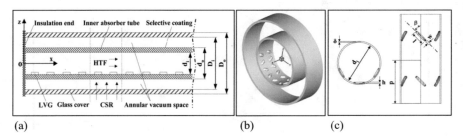

Fig. 3.15 (**a**) The longitudinal cross-section, (**b**) the computational period model and (**c**) parameters of LVG Cheng et al. (2012a, b). (License Number: 4606940725615)

was found that the thermal loss of the UMLVG-PTR reduced by 1.35–12.10% to that of the SAT-PTR within the range studied, and the larger the Reynolds number is, the better comprehensive enhanced thermal effectiveness is. The increase of incident solar radiation increased the average wall temperature and the thermal loss. The UMLVG-PTR has higher overall thermal effectiveness than the smooth PTR, and it has stable performance within a wide range of incident solar radiation.

Islam et al. (2012) performed a 3-D numerical analysis of PTC receiver using CFD-Fluent package as shown in Figs. 3.16 and 3.17. The developed model tackled various physical issues and computational issues associated with PTC system. The physical issues took into consideration non-consistency of the heat flux profile and the temperature profile, and lopsided heating of the working fluid. The most significant

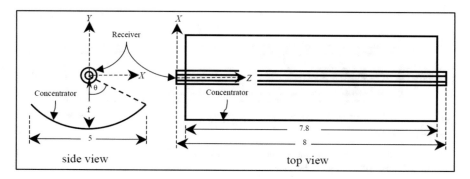

Fig. 3.16 Schematic representation of the PTC (all diemsnions are in "m"). Islam et al. (2012). (License Number: 4606940728823)

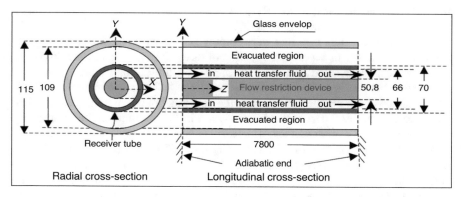

Fig. 3.17 Schematic representation of the PTC absorber (all diemsnions are in "mm"). Islam et al. (2012). (License Number: 4606940728823)

numerical issues incorporate the structure of the calculation area, the turbulence models and the near wall physics. They used effectively and proficiently a basic subroutine program to integrate the receiver's optical information and the boundary conditions into the computational domain simulation. Their simulations affirmed that the optical and the computational simulations of the collector can be practiced autonomously.

Aldali et al. (2013) carried out a CFD numerical analysis for three pitch helical fins with 100, 200 and 400 mm, and an aluminium pipe without fins as shown in Fig. 3.18. The results revealed that the temperature gradient for the pipe without a helical fin was extensively higher contrasted to the pipes with helical fins. It was confirmed that the pipe with 100 mm pitch helical fin demonstrated to be superior to 200- and 400-mm pitch helical fins pipes. Nonetheless, the pressure losses for 100-mm pitch helical fin pipe were in multiple occasions recorded for 400 mm pitch helical fin pipe. Moreover, an aluminium pipe indicated the best outcomes, notwithstanding not having any helical fins. This proposes that the conductivity of the material is significant and has more noteworthy bearing than the nearness of the fins. Thus, the pipes with helical fins can efficiently transfer heat to water than that without fins.

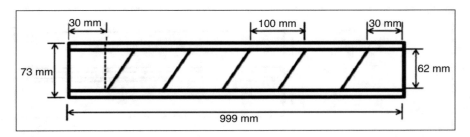

Fig. 3.18 Schematic diagram shows the pipe dimensions with inner helical fins. Aldali et al. (2013). (License Number: 46069407268531)

Wang et al. (2013) investigated the influence of inserted metal foams on the thermal performance of PTC receiver tube as shown in Fig. 3.19. The impacts of metal foams layout, geometrical factor (H) and porosity (u) on the thermal and flow fields were investigated as shown in Table 3.1. The thermophysical properties of metal foams and non-constant heat flux were employed to describe the heat transfer characteristic accurately in superheated section of direct system generation (DSG) system. The results revealed that the "H" impact on the thermal effectiveness was substantial, but this is not the case for the impact of "u" at constant layout and "H" as it is somewhat minimal. Furthermore, the heat flux boundary influences the thermal field essentially. There was 45% reduction in the circumferential temperature difference on PTC's outer surface, which will enormously lessen the thermal stress.

Ghasemi et al. (2013) investigated using CFD FLUENT package the thermal field of PTC having two segmental porous rings as shown in Fig. 3.20. The influence of the distance between the rings on the PTC thermal performance was examined suing Therminol 66 as the heat transfer medium. The outcomes show that the thermal effectiveness of the sun powered parabolic framework was enhanced by the inclusion of permeable two segmental rings. The Nusselt number diminished with increments of the separation between the rings. Therefore, the most extreme Nusselt numbers rely upon the Reynolds number, inner diameter and the separation between permeable two rings. The higher thermal improvement was based on the separations between the two permeable rings Q and d/D being 0.75d and 0.6, respectively.

Ghadirijafarbeigloo et al. (2014) investigated by numerical analysis the thermal effectiveness in PTC receiver pipe fitted with perforated louvered twisted tape (LTT) with various twist ratios (TR) as shown in Fig. 3.21. The results have shown that the thermal and flow fields increased remarkably compared with a pipe having a plain twisted-tape and a smooth pipe. It was found that thermal and flow fields (Nusselt number and friction factor) were higher for LTT tube with a maximum of 150% and 210%, respectively, compared to a smooth pipe. The heat transfer increased with decreasing Re number and TR values, where the optimum results were a TR of 2.67 and Re of 5000.

Too and Benito (2013) conducted a comparative assessment on the thermal and flow fields of solar PTC air absorbers equipped with and without helically coil/wire, twisted tape inserts and dimples as shown in Figs. 3.22 and 3.23. They represented the solar tubular air absorbers using simplified steady-state heat transfer model using CO_2 and He gases. This study revealed that the absorber tubes fitted with

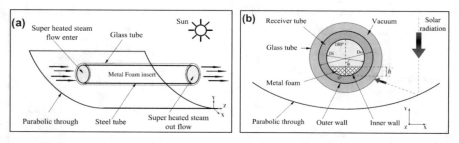

Fig. 3.19 (a) Schematic of PTC inserted with metal foams and (b) PTC cross-section filled with metal foam Wang et al. (2013). (License Number: 4606950012954)

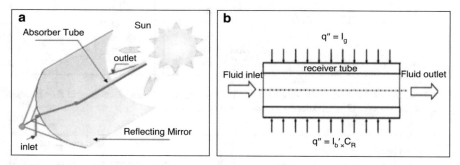

Fig. 3.20 (a) Schematic of PTC system and (b) Cross-section of receiver tube of PTC Ghasemi et al. (2013). (License Number: 4606940728722)

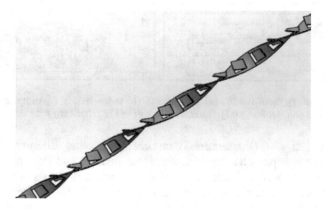

Fig. 3.21 Louvered twisted-tape with perforations Ghadirijafarbeigloo et al. (2014). (License Number: 4611221010240)

dimples shows superior overall enhancement performance without substantial pressure drop penalty compared to that of tubes with coil or tape inserts. It is revealed that using these kinds of surface modifications can minimise the temperature gradient between the wall and airflow. Additionally, the pressure loss could be further diminished with the utilisation of CO_2 and He as working gases.

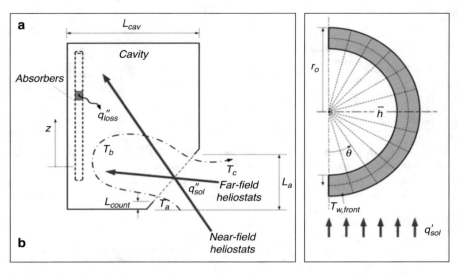

Fig. 3.22 (**a**) A cavity-type solar receiver and (**b**) discretisation of an exial segment of the absrober wall Too and Benito (2013). (License Number: 4611221236456)

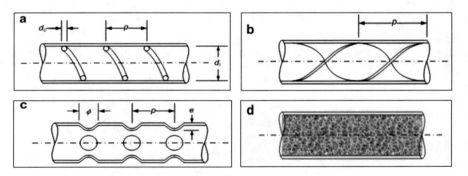

Fig. 3.23 Various types of inserts, (**a**) Helical coil, (**b**) twised insert, (**c**) dimpled tube and (**d**) porous foam Too and Benito (2013). (License Number: 4611221236456)

Waghole et al. (2014) determined experimentally, using different volume fractions of silver nanoparticles concentrations, the thermal and flow fields data in a PTC absorber tube with and without twisted tape inserts as shown in Fig. 3.24. The experiments were conducted using different twist ratios (H/D) and Re number in the laminar range. This study showed that twisted tape insert is an extraordinary guarantee for improving the PTC thermal effectiveness. The experiments show that the Nu number, friction factor and enhancement efficiency were 1.25–2.10 times, 1.0–1.75 times and 135–205%, respectively, over a smooth PTC absorber. It was seen that the silver nanofluid did not produce higher-pressure loss compared to water at a similar twist ratio. New equations for calculating Nu number and friction factor were proposed for both fluids (water and silver).

Song et al. (2014) analysed numerically the impact of including helical screw-tape (HST) inserts and solar incidence angle of PTC absorber tube as shown in Fig. 3.25. Various influence factors including inlet temperature, irradiation level,

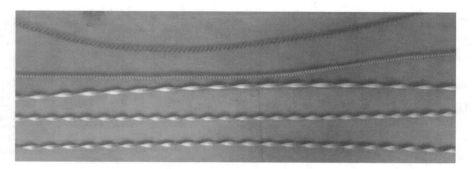

Fig. 3.24 Actual twised tape used in the study of Waghole et al. (2014). (License Number: 4611221237634)

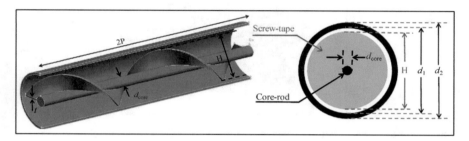

Fig. 3.25 Schematic of absorber tube with HST inserts Song et al. (2014). (License Number: 4611230417674)

helical screw-tape inserts parameters and different flux profiles (transversal angle (β)), were considered. The outcomes revealed that β has a significant effect on the flux profile than longitudinal angle (φ). The relative change of heat loss (Q_{loss}) increased when transversal angle (β) of 11.567 was used. However, its impact diminished with the increment of Re number. It was observed that the HST inserts are an appropriate method of enhancing the thermal effectiveness inside the PTC receiver tube as it decreased the Q_{loss}, T_{max} and ΔT. It was concluded that the HST decreased the heat loss 6 times than the smooth PTR, which increased the pressure loss by 23 times than the smooth PTR.

Mwesigye et al. (2014a, b) tested numerically the usage of centrally placed perforated plate inserts (PPI) in a PTC receiver to examine its thermohydraulic performance as appeared in Fig. 3.26. The examination indicated that he with the utilisation of PPI would provide a range of Re numbers at which the PTC thermal effectiveness is improved. In this range, the thermal efficiency increased from 1.2% to 8%, the Nu number increased from 8% to 133.5% and the friction factor from 1.40 to 95 times contrasted to a plain PTC receiver tube. Relationships for Nusselt number and friction factor were additionally determined and introduced. It was discovered that the altered thermal efficiency is a progressively reasonable assessment instrument since it thinks about the genuine addition in collector execution and the relating increment in pumping loss. The utilisation of PPI provided 52.7% reduction in the entropy generation rate.

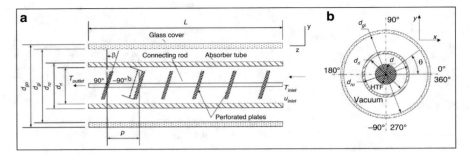

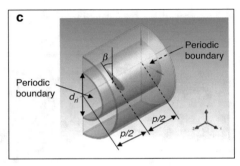

Fig. 3.26 Receiver with PPI, (**a**) longitudinal section, (**b**) cross-section and (**c**) periodic section Mwesigye et al. (2014a, b). (License Number: 4611231005708)

Syed Jafar and Sivaraman (2014) investigated experimentally the impact of using a PTC receiver equipped with nail twisted tape (NTT) of two different twist ratios. They studied the effect of Al_2O_3/water nanofluid with 0.1%, and 0.3% particle volume concentration on the thermal and flow fields as shown in Fig. 3.27. The tests were conducted using indoor simulation under uniform heat flux condition at Re number range of 710–2130. It was concluded that the NTT absorber with nanofluids can remarkably enhance the PTC thermal effectiveness. The results revealed that the friction factor increased with NNT due to high swirl flow.

Sadaghiyani et al. (2014) performed numerical study using finite volume method for thermal and flow fields of LS-2 PTC as shown in Fig. 3.28. Mont-Carlo statistical technique written by MATLAB was used to determine the variation of solar heat flux around the receiver tube. The results show that the natural convection was dominant over the forced convection for r* (plug diameter) < 0.6 m. However, the mixed convection becomes the predominant instrument for r* > 0.6 m. In addition, the friction factor decreased, and the Nu number was minimum at r* = 0.6 m when the plug diameter increased. The outlet temperature is increased with the increment of the tube thermal conductivity. However, this impressibility is negligible at high thermal conductivities.

Mwesigye et al. (2015) extended his previous work and used multi-objective and thermodynamic optimisation methods to study the thermal effectiveness of a PTC receiver fitted with perforated plate inserts (PPI) as shown in Fig. 3.26. The optimisation included several dimensionless perforated plate geometrical parameters such

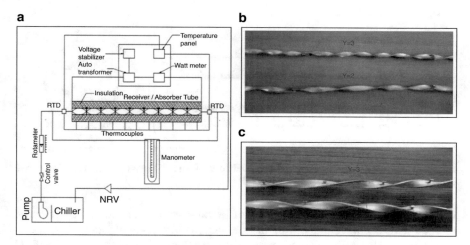

Fig. 3.27 (**a**) Experimental setup diagram, (**b**) Plain twisted tapes (PTT) and (**c**) Nail twisted tapes (NTT) Syed Jafar and Sivaraman (2014). (License Number: 4611231006329)

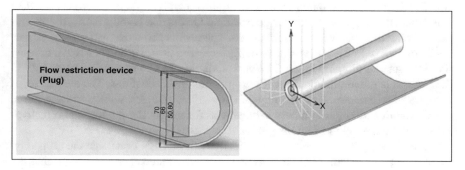

Fig. 3.28 Schematic diagram of PTC and the cross section of LS-2 absorber tube Sadaghiyani et al. (2014). (License Number: 4611231006330)

as its inclination angle, plate diameter and plate spacing. The multi-objective opti-misation was conducted via the combined utilisation of CFD, design of experi-ments, response surface methodology and Genetic Algorithm-II. The results revealed that the entropy generation rate decreased as the angle of orientation increased. It was also concluded that the optimal Re number decreased as the plate size increased and plate spacing decreased with an entropy generation rate decre-ment by 53%.

Mwesigye et al. (2015) performed a thermodynamic analysis and entropy gen-eration for a PTC receiver tube using a synthetic oil-Al_2O_3 nanofluid. The PTC system has of 80° rim angle and 86 proportion ratios as shown in Fig. 3.29. The Re number ranged between 3560 and 1,151,000 with nanoparticle concentration in the range of 0–8%. The results revealed that the thermal efficiency enhanced by up to 7.6% when using nanofluids. It was pointed out that at each inlet temperature and volume fraction there is an optimum value of Reynolds number where minimum

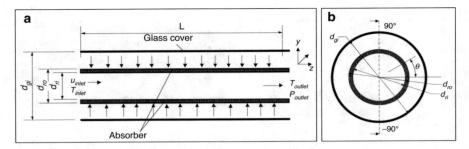

Fig. 3.29 The PTC receiver used: (**a**) longitudinal view and (**b**) cross-section view Mwesigye et al. (2015). (License Number: 4611231005708)

entropy is occurred. There was a value of Re number beyond which the utilisation of nanofluids is unfavourable from the thermodynamics point of view.

Huang et al. (2015) studied numerically the thermal effectiveness of a PTC receiver equipped with helical fins, protrusions and dimples as shown in Fig. 3.30. The outcomes revealed that the thermal performance was superior for receiver tubes fitted with dimples compared with tubes equipped with either protrusions or helical fins. Then, the impacts of dimples geometrical factors and layouts on the thermohydraulic effectiveness were further studied. The results show that as Re number varies from 1×10^4 to 2×10^4, the value of f/f_0 increased 56–77% and Nu/Nu_0 increased 44–64% for the dimpled tubes. The corresponding PEC were varied from 1.23 to 1.37 which is greater than tubes with protrusions and helical fins. The evaluation of heat transfer enhancement techniques shows that dimples with narrower pitch, deeper depth and larger numbers is beneficial for improving the thermal effectiveness whilst other layouts had insignificant impact.

Chang et al. (2015) examined numerically with FLUENT package the utilisation of twisted tapes inserts (TTI) in a molten salt solar receiver tube to determine the turbulent thermal effectiveness as shown in Fig. 3.31. Various parameters of twisted tapes were studied such as the clearance ratios © and twist ratios (TR) on the thermal and flow fields under non-uniform heat flux. The outcomes indicated that the TTI remarkably enhanced the consistency of temperature distribution of tube wall and molten salt. The decreases of C and TR enhanced the thermal effectiveness effectively and increased the friction factor.

Lu et al. (2015) analysed theoretically the utilisation of spirally grooved pipe of solar receiver to enhance heat transfer performances as shown in Figs. 3.32 and 3.33. Based on the obtained outcomes, when both the flow velocity and groove height increased, the absorption efficiency increased, whilst the wall temperature decreased. It was found that the heat absorption efficiency of solar receiver fitted with spirally grooved pipe having e/d = 0.0475 increased about 0.7%, and the optimum bulk fluid temperature increased to 31.1 °C compared with a solar heat receiver with a smooth pipe. It was concluded that spirally grooved pipe can be an exceptionally viable route for thermal absorption improvement, and it can likewise produce high temperature of molten salt.

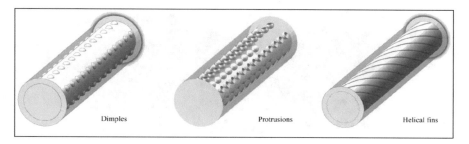

Fig. 3.30 The schematic diagram of receiver tube with dimples, protrusions and helical fins Huang et al. (2015). (License Number: 4611231005719)

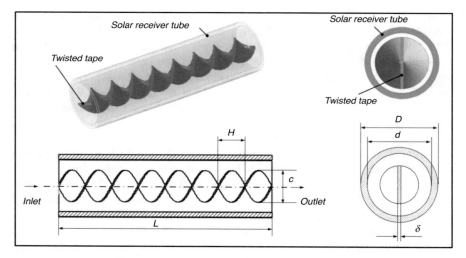

Fig. 3.31 The geometry of a PTC receiver tube with TTI Chang et al. (2015). (License Number: 4611231005688)

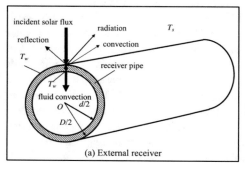

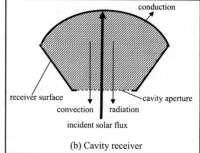

Fig. 3.32 Basic heat absorption model of solar receiver Lu et al. (2015). (License Number: 4612900851490)

Fig. 3.33 Experimental installation and spirally grooved tube Lu et al. (2015). (License Number: 4612900851490)

Zhu et al. (2015) evaluated experimentally a coil type solar dish receiver at real conditions to study the energy and exergy performance as shown in Figs. 3.34 and 3.35. The results have shown that the efficiency is more than 70% and it could reach to about 82%, while the efficiency at steady state was remained about 80%. The optimum energy rate can be 21.3 kW while the exergy rate is about 8.8 kW. Besides, the maximum energy efficiency and exergy efficiency are 82% and 28%, respectively. It was concluded that the impact of the exergy factor increased as the temperature difference increased. A lower exergy factor value of 0.3 is obtained although the temperature difference is high at 250 K.

Reddy et al. (2015) carried out experiments on a 15 m² solar PTC equipped with porous disc receiver as shown in Fig. 3.36 using six different receiver configurations to evaluate their performance as shown in Figs. 3.37 and 3.38. The experiments were done for different weather conditions and a flow rate in the range of 100–1000 L/h. The results revealed that the time constant for different PTC receiver configurations was in the range of 70–260 s. The collector acceptance angle is found to be 0.58° and 0.68°, respectively, for both unshielded receiver (UR) and shielded receiver (SR). The collector efficiencies were in the range of 63.9–66.66% with pumping loss of 0.05 W/m and the heat losses were in the range of 455–1732 W/m². It was inferred that the porous disc receiver remarkably enhanced the PTC performance, and it thus can be utilised viably for process heat applications.

Diwan and Soni (2015) utilised PTC absorber tube equipped with wire-coils inserts to examine numerically the thermal and hydraulic characteristics using water as a working medium. The numerical simulations were conducted with the aid of COMSOL Multiphysics 4.4 for different Reynolds numbers and various wire-coils insert pitch values. The results revealed that the Nu number increased by 104–330% and the maximum pressure loss ranged between 55.23 and 1311.79 Pa in case of wire-coils with pitch 8 mm. It was concluded that for better thermal effectiveness it is preferable to use wire-coils inserts with pitch value in the range of 6–8 mm and more than 8 mm when working with low and high Re numbers, respectively.

Mwesigye et al. (2016a, b) utilised wall-detached twisted tape inserts in a PTC absorber tube to investigate numerically the thermal effectiveness based on a FVM as shown in Fig. 3.39. It was found that when the twist ratio reduced and the width

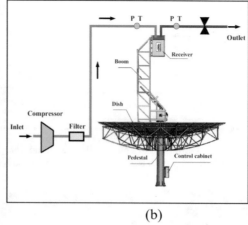

(a)   (b)

Fig. 3.34 (**a**) Parabolic dish and (**b**) Schematic diagram of the receiver test arrangement Zhu et al. (2015). (License Number: 4612901138085)

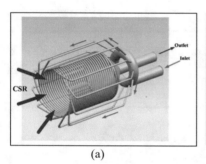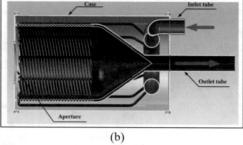

(a)   (b)

Fig. 3.35 (**a**) 3D model of solar receiver and (**b**) Solar receiver section view Zhu et al. (2015). (License Number: 4612901138085)

Fig. 3.36 Solar PTC system Reddy et al. (2015). (License Number: 4612901352313)

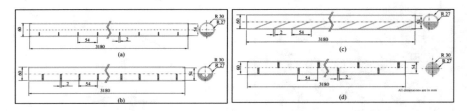

Fig. 3.37 Various porous disc receiver configurations, (**a**) bottom porous, (**b**) inclinded, (**c**) vertical and (**d**) top porous Reddy et al. (2015). (License Number: 4612901352313)

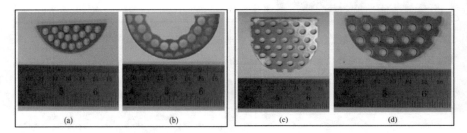

Fig. 3.38 Photograph of porous disc for, (**a**) bootom, (**b**) inclind, (**c**) vertical and (**d**) top porous Reddy et al. (2015). (License Number: 4612901352313)

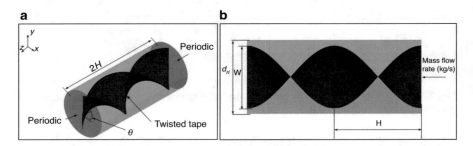

Fig. 3.39 Receiver's absorber tube geometry: (**a**) full view and (**b**) section view Mwesigye et al. (2016a, b). (License Number: 4611231005934)

ratio increased, both the thermal and flow fields increased in the range of 1.05–2.69 times whilst the fluid friction was in the range of 1.6–14.5 compared to a plain receiver pipe. The results have shown significant increase in both thermal efficiency and heat transfer performance of about 10%, 169%, respectively, and reduction in absorber's circumferential temperature up to 68% over a plain receiver pipe. The entropy generation is minimal for each twist ratio and width ratio at the optimum Re number. Empirical equations for thermal and hydraulic performance were also developed.

Nems and Kasperski (2016) presented internal multiple-fin array in a concentrated solar air-heater to study its effect experimentally from the energy validation perspective as shown in Figs. 3.40 and 3.41. They developed a new solar air heater working at high temperature to convert solar energy to heat for space heating in

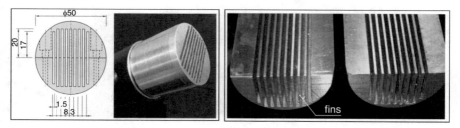

Fig. 3.40 Absorber with internal multiple-fin array Nems and Kasperski (2016). (License Number: 4612940949527)

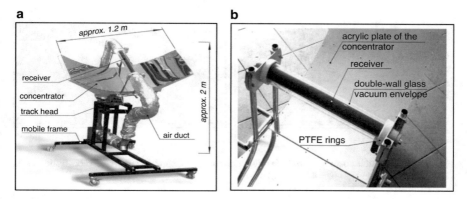

Fig. 3.41 PTC with cylindrical receiver as an air heater: (**a**) full experimental setup and (**b**) absorber with its details Nems and Kasperski (2016). (License Number: 4612940949527)

Poland's weather conditions. The objective of the study was to verify the previously created mathematical model of heat transfer processes. The collector's performance was analysed experimentally against the reduced temperature difference. The difference between the model and experimental results shows a relatively small deviation between them. Deviation of thermal efficiency does not exceed ±5.4%. However, the proposed absorber design was a prototype with some technological weaknesses. It was not industrially produced, as it was the first stage of the research programme. Generally, the two weak sides of the solution were: lack of tightness of the absorber and lack of vacuum control of the glass casing. It was suggested that the absorber should be longitudinally welded, covered with selective oxide layer and mounted inside a single glass pipe. It would enable to obtain a vacuum from the glass cover, which would allow to increase the thermal efficiency.

Bellos et al. (2016a, b) utilised a dimpled with sine geometry PTC absorber tube to study its thermal efficiency enhancement as shown in Figs. 3.42 and 3.43. Different working mediums were used thermal oil, thermal oil-Al_2O_3 nanofluid and pressurised water to examine its efficacy on the overall thermal effectiveness. The outcomes revealed that the thermal oil-Al_2O_3 nanofluid produced 4.25% improvement of the mean efficiency while 6.34% is produced using pressurised water. The results revealed that the mean efficiency enhanced by 4.55% compared to plain tube

Fig. 3.42 Examined PTC
model using solidworks
Bellos et al. (2016a, b).
(License Number:
4612941218888)

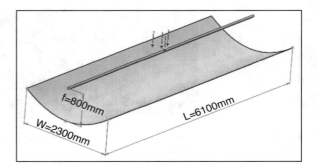

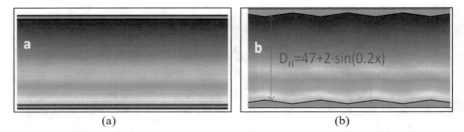

Fig. 3.43 Examined absorber tubes: (**a**) cylinderical tube and (**b**) sine shape tube Bellos et al. (2016a, b). (License Number: 4612941218888)

geometry. Higher fluid temperature levels produced an increase in the efficiency. However, the use of a wavy surface produced negative outcome of high-pressure losses which needs to be considered.

ZhangJing et al. (2016) utilised porous insert in a solar PTC receiver tube under non-uniform heat flux condition to investigate the thermal enhancement as shown in Figs. 3.44, 3.45, and 3.46. They used an optimisation method, which couples genetic algorithm (GA) and CFD to optimise the layout of porous insert. Three thermal effectiveness evaluation criterions were used such as synergy angle, entransy dissipation and exergy loss. It was observed that the thermal effectiveness of porous receiver insert is always higher than that of the plain receiver. With the use of some materials of high thermal conductivity (Cu, Al and SiC), the solar-to-thermal energy conversion efficiency of GA porous receiver can reach 68%, which is higher than the referenced porous insert receiver is. It was noticed that the energy transfer and exergy loss rate had similar influences on the thermal irreversibility.

Fuqiang et al. (2016a, b) used asymmetric outward convex corrugated tube (AOCCT) tube of PTC receiver to increase its reliability and overall thermal effectiveness as shown in Figs. 3.47 and 3.48. The PTC thermal effectiveness and strain was studied by developing an optical-thermal-structural coupled method. It was indicated that the usage of AOCCT can effectively improve the thermal effectiveness and reduce the thermal strain. The results revealed that the maximum augmentation of overall thermal effectiveness factor and restrain of von-Mises thermal strain are 148% and 26.8%, respectively, with the utilisation of AOCCT as receiver.

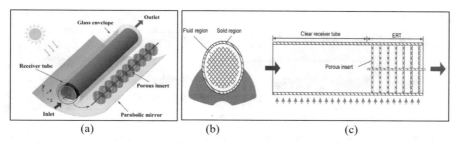

Fig. 3.44 (**a**) PTC segment diagram, (**b**) Physical model for simulation transverse cross-section and (**c**) Longitudinal cross-section ZhangJing et al. (2016). (License Number: 4612950191523)

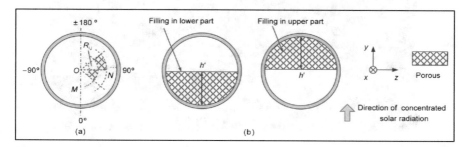

Fig. 3.45 Schematic of some layouts of porous insert, (**a**) GA and (**b**) reference ZhangJing et al. (2016). (License Number: 4612950191523)

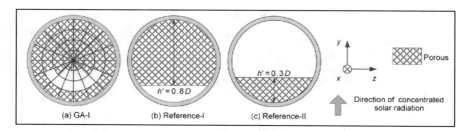

Fig. 3.46 Some configurations of porous inserts suitable for ERT ZhangJing et al. (2016). (License Number: 4612950191523)

Fig. 3.47 AOCCT as PTC receiver Fuqiang et al. (2016a, b). (License Number: 4612950656289)

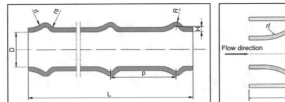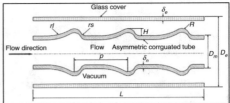

Fig. 3.48 Diagram of AOCCT used for PTC tube Fuqiang et al. (2016a, b). (License Number: 4612950656289)

Fig. 3.49 Experimental setup shows PTC-90-II Jaramillo et al. (2016). (License Number: 4612950818541)

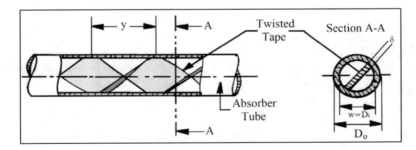

Fig. 3.50 Diagram of twisted tape insert used Jaramillo et al. (2016). (License Number: 4612950818541)

Jaramillo et al. (2016) used twisted tape technique (TTT) to study numerically and experimentally, as shown in Figs. 3.49 and 3.50, the heat transfer augmentation in the receiver tube for low enthalpy processes by thermodynamics laws. The outcomes revealed that the Nusselt number, removal factor, friction factor and thermal efficiency increased when both the twist ratio (y/w) and Re number reduced compared with the plain receiver tube. However, these quantities do not show an improvement when the twist ratio (y/w) increased. It was concluded that TTT is an effective way to improve PTC thermal effectiveness only under certain conditions. These optimal conditions correspond to a system having a twist ratio close to 1, and

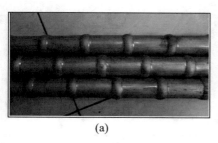
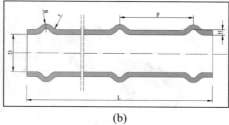

(a)   (b)

Fig. 3.51 (**a**) Outward convex transverse corrugated tubes and (**b**) Corrugated tube used Fuqiang et al. (2016a, b). (License Number: 4612950656289)

low flow rates in the order of 1 L/min. It was clear that the thermal efficiency increased as the volumetric flowrate increased.

Fuqiang et al. (2016a, b) used a symmetric outward convex corrugated tube (SOCCT) design for PTC receivers to increase their thermal performance and reliability as shown in Fig. 3.51. To analyse the thermal deformation of the PTC tube an optical-thermal-structural model was developed. The numerical outcomes revealed that the utilisation of SOCCT effectively enhanced the thermal effectiveness and reduced the thermal strain. It was observed that the heat transfer factor raised by 8.4%, and the maximum thermal strain of metal tube reduced by 13.1% when SOCCT is utilised. In addition, empirical correlations were proposed to find the effective Nusselt number and pressure drop in a PTC.

Benabderrahmane et al. (2016) investigated numerically the thermal effectiveness of a PTR equipped with longitudinal fins inserts (LFI) and various types of nanofluid as shown in Fig. 3.52. The Monte Carlo ray tracing technique was used to obtain the non-uniform heat flux profile. A remarkable thermal enhancement was attained when Reynolds number varied from 2.57×10^4 to 2.57×10^5, and the Nusselt number increased from 1.3 to 1.8 times. It was found that the metallic nanoparticles significantly enhanced the heat transfer than other nanoparticle types. The friction factor for absorber with fins varied from 1.6 to 1.85 than plain tube. The geometric parameters of the fins have a remarkable effect in heat transfer improvement. At similar condition, using nanofluid in absorber with fins insert offer higher heat transfer performance and higher thermo-hydraulic performance than smooth tube with base fluid.

Chang et al. (2017) utilised concentric and eccentric pipe inserts in a PTR with molten salt as heat transfer medium to examine its performance numerically as shown in Fig. 3.53. A 3D simulation model was established, and the combination of a MCRT method and FLUENT software was utilised to obtain the non-uniform heat flux. The results show that the introduction of concentric pipe inserts for PTR augmented the overall thermal effectiveness gradually with maximum augmentation factor of A3 (D = 0.04 m) is about 164%. The maximum temperature decrease by A3 can reach more than 19.1 K and 2.7 K in the absorber tube and the molten salt, respectively. The eccentric pipe inserts of B3 (D = 0.04 m and e = 0.01 m) performs significantly better than the concentric tube inserts for its excellent performance in decreasing the maximum temperature of absorber tube and molten salt. It enhanced the overall thermal effectiveness dramatically with maximum augmentation factor

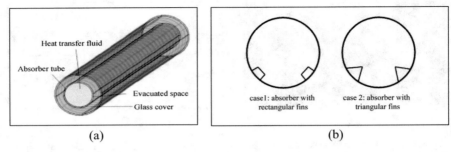

(a)   (b)

Fig. 3.52 (**a**) Computational domain diagram and (**b**) Cross-section of absorber tube with LFI Benabderrahmane et al. (2016). (License Number: 4612960050303)

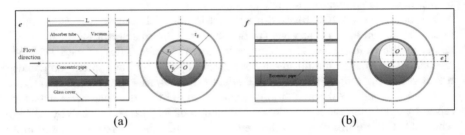

(a)   (b)

Fig. 3.53 (**a**) PTC system with concentric pipe inserts and (**b**) PTC system with eccentric pipe inserts Chang et al. (2017). (License Number: 4611231006837)

of B3 is about 165%. The maximum temperature decreased more than 20 K and 2 K in the absorber tube and the molten salt, respectively.

Ghasemi and Ranjbar (2017a, b) simulated 3D turbulent flow of Syltherm oil in a PTC absorber tube having porous rings on the thermo-hydraulic characteristics as shown in Fig. 3.54. The numerical simulation was carried out using CFD with the aid of ANSYS commercial software. The impact of distance between rings and the rings inner diameter on the thermal effectiveness of the PTC was evaluated. The outcomes have shown that the thermal effectiveness is significantly enhanced by utilising the porous rings. It was found that the thermal characteristics increased by decreasing the distance between the porous rings, but the Nusselt number is reduced by increasing the rings inner diameter as shown in Fig. 3.55.

Bortolato et al. (2017) tested nanofluids experimentally in a full-scale direct absorber-concentrating collector as shown in Fig. 3.56. The nanofluid used was single wall carbon nanohorns (SWCHNs) suspended in distilled water with volume fraction of 0.02 g/L. A direct absorption receiver was constructed with a flat geometry and installed on an asymmetric PTC using 100 kW/m^2 of solar flux. The results show that the receiver thermal efficiency initially slightly changed and reached 87% and then after 8 h of exposure it continuously decreased down to 69%. Spectrophotometric analysis on bulk nanofluid samples examined at various times shows that the SWCNHs concentration in water was unstable. This was due to coalescence and precipitation of the aggregates and the polymeric surfactant rapid

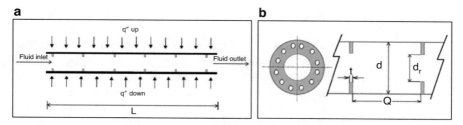

Fig. 3.54 (**a**) Diagram of PTC absorber tube with porous rings and (**b**) Cross-section of porous rings Ghasemi and Ranjbar (2017a, b). (License Number: 4613491213097)

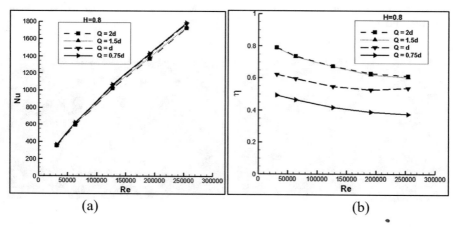

Fig. 3.55 (**a**) Nusselt number for different distance between porous rings and (**b**) Thermal performance factor for different distance between porous rings Ghasemi and Ranjbar (2017a, b). (License Number: 4613491213097)

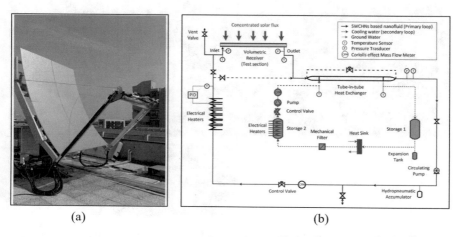

Fig. 3.56 (**a**) Prototype of the asymmertical linear PTC and (**b**) Schematic of the experimental test rig Bortolato et al. (2017). (License Number: 4613500036260)

degradation. The fluid circulation may contribute to the nanoparticles aggregation as shown in Fig. 3.57.

Rawani et al. (2017) used serrated twisted tape insert (STTI) in a PTC absorber tube to investigate analytically its thermal performance as shown in Fig. 3.58. The thermal equations for fully developed flow were developed to study the variation of entropy generation, exergy and thermal efficiencies and fluid temperature distribution. The temperature rise of fluid for system performance evaluation under specified conditions was calculated by developing a computer program using C^{++} language. The results have shown that the STTIs are a good technique for improving the PTC performance. The results also revealed that for twist ratios x = 1, the Nusselt number was shown higher 4.38 and 3.51 times over a plain PTC with corresponding thermal efficiency of 15.7% and 5.41%, respectively, and exergy efficiency of 12.10% and 12.62%.

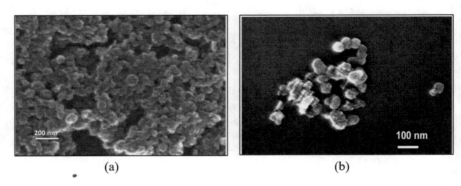

(a)   (b)

Fig. 3.57 (**a**) FE-SEM micrograph of SWCHNs powder and (**b**) SEM micrograph of 15 min homogenised sample Bortolato et al. (2017). (License Number: 4613500036260)

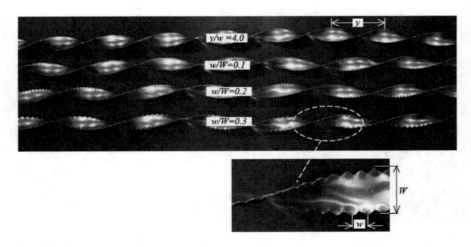

Fig. 3.58 Schematic diagram of serrated twisted tape inserts Rawani et al. (2017)

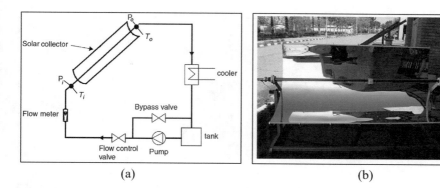

Fig. 3.59 (**a**) The experimental test system used and (**b**) PTC physical model Jamal-Abad et al. (2017). (License Number: 4613500396915)

Fig. 3.60 Copper foam as a porous media Jamal-Abad et al. (2017). (License Number: 4613500396915)

Jamal-Abad et al. (2017) investigated practically the thermal effectiveness of a PTC filled with copper foam as shown in Figs. 3.59 and 3.60. The experiments were conducted for flow rates (0.5–1.5 L/min), the copper foam porosity and pore density were 0.9 and 30 PPI (pores per inch), respectively. The results revealed that the thermal effectiveness was improved, and the loss factor reduced by 45% when using copper foam.

Pavlovic et al. (2017a, b) developed a thermal model for a dish reflector and spiral absorber of solar collector using Engineering Equation Solver (EES) as shown in Figs. 3.61 and 3.62. The commercial software OptisWorks was utilised to perform the numerical simulations. They used energy and exergy analyses to compare the thermal effectiveness of three fluids, i.e., water, therminol VP-1 and air. The results revealed that the PTC's thermal efficiency is only about 34% due to the low

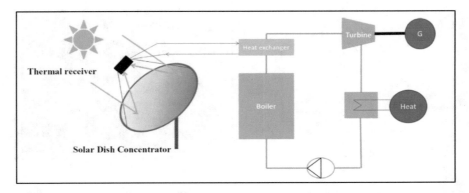

Fig. 3.61 Cogeneration plant with solar dish concentrating collector Pavlovic et al. (2017a, b). (License Number: 4613500612072)

(a)   (b)

Fig. 3.62 (a) The examined solar PTC system and (b) Cross-section of solar receiver Pavlovic et al. (2017a, b). (License Number: 4613500612072)

optical efficiency. The exergetic results also affirmed that water was the most suitable working fluid compared to other fluids, as it was capable to effectively operate at low temperatures, while the thermal oil was the best at higher temperatures. They suggested using conical shape open receiver having helical tube for improving the receiver's optical effectiveness which can efficiently absorb an average flux value of 2.6×10^5 W/m^2.

Bellos et al. (2017a, b, c, d, e) utilised 12 different internally longitudinal fins in an absorbers of PTC module under different operating conditions as shown in Fig. 3.63. SolidWorks Flow Simulation was used as the simulation analysis tool and the model results were validated against the literature results. Various operational conditions were tested with the inlet temperature range of 300–600 K and the flow rate range of 50–250 L/min. It was found that both fin thickness and greater length lead to higher thermal effectiveness and higher-pressure loss. It was concluded that overall optimum absorber had fin length of 10 mm and fin thickness of 2 mm. It was observed that the thermal efficiency and the Nusselt number were improved by 0.82%, and 65.8%, respectively. Empirical correlations for Nusselt number and

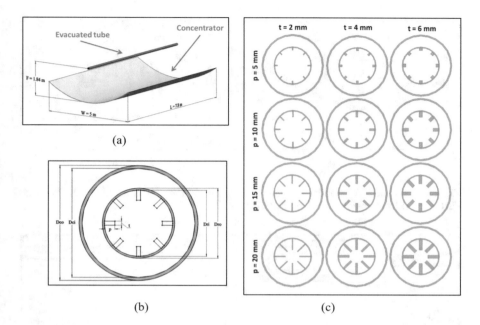

Fig. 3.63 (a) The PTC examined module, (b) Finned absorber cross section and (c) Twelve tested internally fined absorbers Bellos et al. (2017a, b). (License Number: 4613500886237)

friction factor were developed and it was affirmed that both are applicable for smooth and finned absorbers.

Bellos et al. (2017a, b, c, d, e) continued his previous work and investigated the thermal effectiveness of internally finned absorbers as shown in Fig. 3.64. They particularly examined internal fins with thicknesses of 2, 4 and 6 mm, whilst their lengths are 5, 10, 15 and 20 mm. They used 600 K as the inlet temperature and 150 L/min as the flow rate because these conditions provided an adequate thermal performance based on their previous study. The results have shown that the influence of the length is more exceptional than the thickness. It was found that the optimum absorber had 20 mm of length and 4 mm of thickness as shown in Fig. 3.65. Accordingly, the thermal efficiency and the thermal enhancement factor were increased by 1.27% and 1.483, respectively.

Bellos et al. (2017a, b, c, d, e) examined the use of internal longitudinal fins in PTC utilising different gas types (Air, helium and carbon dioxide) as shown in Fig. 3.66. A commercial software Solidworks Flow Simulation was used to perform the simulations. The collector was checked from the energy and exergy point of views to determine the useful output and the pumping loss. It was found that the fin of 10 mm was the best, and helium was the most efficient working fluid from the exergetic analysis, as it performs better until 290°C. The CO_2 was the best working fluid at higher temperatures as shown in Fig. 3.67. Moreover, the optimum flow rate for other working fluids was 0.015 kg/s and 0.03 kg/s for helium.

Zhu et al. (2017) [128] established a comprehensive numerical model using CFD to study the flow and thermal fields inside an absorber tube of PTR equipped with wavy-tape insert using Syltherm-800 as shown in Fig. 3.68. It was found that

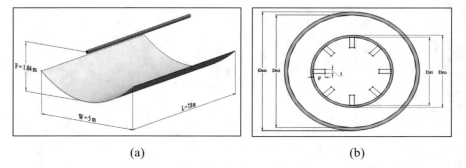

(a)   (b)

Fig. 3.64 (**a**) The PTC examined module and (**b**) Finned absorber cross section Bellos et al. (2017a, b). (License Number: 4613510216043)

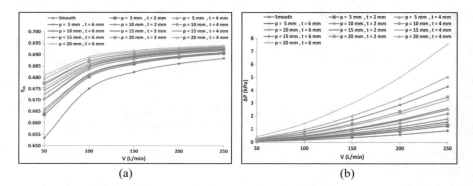

(a)   (b)

Fig. 3.65 (**a**) Thermal efficiency and (**b**) Pressure drop at inlet temperature of 600 K Bellos et al. (2017a, b, c, d, e). (License Number: 4613510216043)

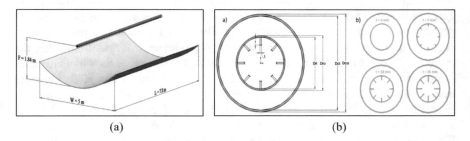

(a)   (b)

Fig. 3.66 (**a**) The PTC examined module and (**b**) The finned absorber cross section with different fins dimensions Bellos et al. (2017a, b, c, d, e). (License Number: 4613510442895)

wavy-tape produced high thermal augmentation and lower tube temperature and heat loss. However, the wavy-tape increased the PTR thermal stress and deformation with increased pressure loss. The Nusselt number and friction factor of the wavy PTR increased by 261–310% and 382–405%, respectively. The wavy-tape produced a performance evaluation index more than 2.11 and heat loss reduction by 17.5–33.1% with entropy generation rate reduction by 30.2–81.8%.

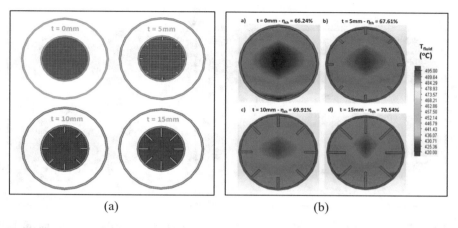

Fig. 3.67 (**a**) The mesh generation for four cases and (**b**) Temperature profile at different thicknesses using air Bellos et al. (2017a, b, c, d, e). (License Number: 4613510442895)

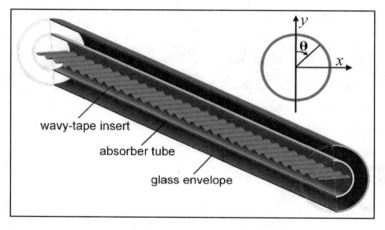

Fig. 3.68 Schematic diagram of the PTR equipped with wavy-tape insert Zhu et al. (2017). (License Number: 4613511068043)

Bellos et al. (2017a, b, c, d, e)used energy and exergy analyses to identify the enhancement percentage of a PTC equipped with internal fins using Solidworks Flow Simulation as shown in Fig. 3.69a. Carbon dioxide was used as a working gas to examine the system's performance at high temperature levels. The influences of flow rate and longitudinal fins on the PTC's performance were studied as shown in Fig. 3.69b. The results revealed that the thermal performance increased when using higher fin length. It was found that 10 mm was the optimum fin length with 45.95% exergetic efficiency. It was also concluded that the pressure losses along the collector was the best criteria for assessing solar collectors working with gases.

Xiangtao et al. (2017) used pin fin arrays insert (PFAI) in a PTC receiver tube to enhance increase the thermal effectiveness as shown in Fig. 3.70. The combination of Monte Carlo ray tracing method (MCRT) coupled with Finite Volume Method (FVM) was adopted to perform the simulations. The numerical results indicated that

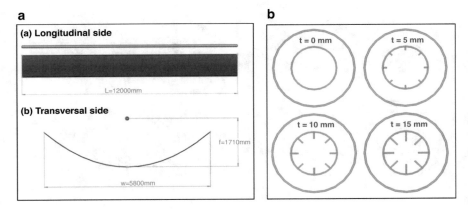

Fig. 3.69 (a) The logitudinal side and the transverse side of the examined PTC module and (b) the examined absorber with and without internals fins Bellos et al. (2017a, b)

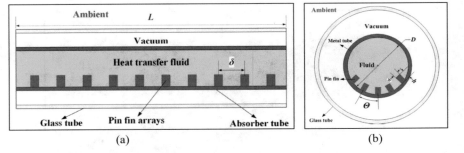

Fig. 3.70 The schematic diagram of PTC tube equipped with PFAI, (a) longitudinal section and (b) cross-section Xiangtao et al. (2017). (License Number: 461351138278)

the introduction of PFAI in PTC receiver can significantly improve the thermal effectiveness. The thermal effectiveness factor and the Nusselt number increased up to 12.0% and 9.0%, respectively, with the unitisation of PFAI. The optimum conditions for higher thermal effectiveness were attained.

Huang et al. (2017) studied numerically a 3D turbulent mixed heat transfer in dimpled tubes of PTC tube as shown in Fig. 3.71. The impacts of outer wall heat flux profiles and dimple depth on the thermal and flow fields were analysed. The outcomes revealed that the thermal and flow performance in dimpled receiver tubes subjected to non-uniform heat flux (NUHF) were larger than those subjected to uniform heat flux (UHF). It was found that the deep dimples (d/Di = 0.875) were performing much better than the shallow dimples (d/Di = 0.125) at similar Grashof number. It was concluded the impacts of impingement, reattachment and vortex shedding were weakened in the top dimple because of the buoyancy force influence, while the flow was accelerated, and the vortex moved backward in the bottom dimple. The higher thermal performance was achieved because of the high reattachment flow, high impingement and vortex shedding.

Chang et al. (2018) analysed the reliability and overall convective heat transfer of molten salt flow in a PTC tube with the usage of concentric rod and eccentric rod

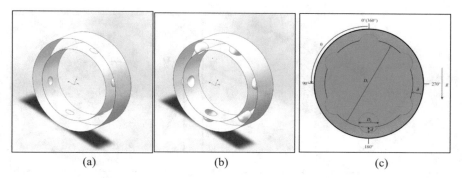

Fig. 3.71 The dimpled tube representation, (**a**) shallow dimples, (**b**) deep dimples and (**c**) parameters of dimpled receivers Huang et al. (2017). (License Number: 4613520033551)

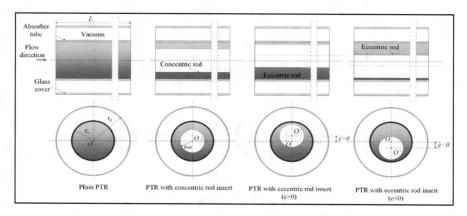

Fig. 3.72 The geometry and grid of the plain PTC with concentric rod and eccentric rod inserts Chang et al. (2018). (License Number: 4613530302030)

as turbulators as shown in Fig. 3.72. The result showed that both concentric and eccentric rod inserts enhanced the thermal effectiveness effectively. For a PTC with a concentric rod insert, the Nusselt number is about 1.10–7.42 times over a plain PTC when the dimensionless diameter (B) increased. The thermal effectiveness factor has a remarkable decrease with Reynolds number increment when B is larger than 0.8. For an eccentric rod insert, the performance effectiveness decreased when Reynolds number increased under a certain dimensionless eccentricity (H). The performance effectiveness decreased from 1.84 to 1.68 times over a plain PTC when H is 0.8. In addition, the absorber tube's optimum temperature decreased dramatically when B and H increased, which eventually reduced the PTC thermal deflection and increased its reliability.

Bellos et al. (2018a, b, c) used a star shape insert for enhancing the thermal performance of PTC. SolidWorks Simulation Studio was used to conduct the the analysis and comapred against a varified model. A total of 16 cases were examined with a fin length range of 15–30 mm and fin thickness range of 2–5 mm as shown in Fig. 3.73. The results revealed that the Nusselt number enhancement is up to 60% with thermal losses decrement up to 14%. It was inferred that the thickness of 5 mm

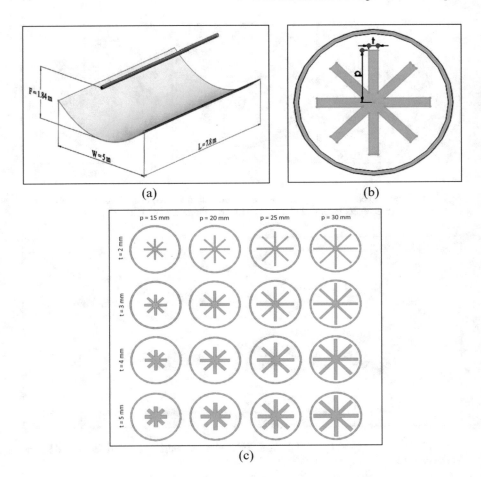

Fig. 3.73 (**a**) The examined PTC, (**b**) Star insert shape and (**c**) Star flow insert cases Bellos et al. (2018a, b,c). (License Number: 4613540771277)

and fin length range of 20–30 mm were the optimum paramters for star insert fins. It was also pointed out that the utilisation of flow inserts would decrease the absorber circumferential temperature range and this would in turns decrease the failure danger caused by the thermal stresses. It was observed that about 15% maximum decreament in the temperature range could be achieved with star insert compared to smooth absorber tube.

Bellos et al. (2018a, b, c) exploited the results of their previous studies and investigated the internal fins best number and location in a PTC absorber using multi-objective procedure as shown in Fig. 3.74. The analysis was conducted with SolidWorks Flow Simulation using rectangular fins. It was concluded that the internal fins should be located in the absorber tube's lower part where the heat flux is highly concentrated. The results affirmed that the optimum case was with an absorber tube having three fins in its lower part ($\beta = 0°$, $\beta = 45°$ and $\beta = 315°$) with thermal efficiency of 68.95% higher than the efficiency of the case with one fin.

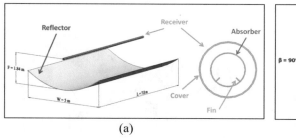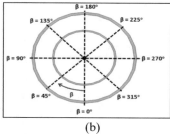

(a)   (b)

Fig. 3.74 (**a**) The examined PTC with its internally finned absorber and (**b**) Internal fins locations in the absorber tube Bellos et al. (2018a, b, c). (License Number: 4613540976165)

Bitam et al. (2018) conducted a 3D numerical model to investigate thermal effectiveness in a PTC having conventional straight and smooth tube (CSST) receiver is replaced by a newly designed with a S-curved/sinusoidal tube receiver using synthetic oil as shown in detailed in Fig. 3.75. It was noticed that the mean Nusselt number increased by 45–63%, while the friction coefficient increased by less than 40.8%, which leads to a maximum performance evaluation criterion about 135%. The maximum CTD of the PTR S-curved tube decreased below 35 K for all cases studied which in turns leads to both thermal stresses and heat losses reduction. The thermal augmentation overcomes the corresponding pressure loss by a factor less than 1.35.

Tripathy et al. (2018) performed thermal-fluid and structural analyses of absorber tube used in PTC with different materials using computational approach and utilising Therminol VP1 as shown in Figs. 3.76, 3.77, and 3.78. Steel, copper, aluminium and bimetallic (Cu-Fe) and tetralayered laminate (Cu-Al-SiC-Fe) were used. It was found that there is no influence for the absorber's material change on the heat transfer, but it does have a remarkable influence on the bending of the tu©ause of the thermal expansion and self-weight. The results revealed that the steel tube and copper tube absorbers have poor circumferential temperature distributions and higher self-weight and lower mechanical strength, respectively. Thus, it was suggested to use a bimetallic tube, which decreases the deflection by 7–15% compared to steel. Besides, tetra-layered laminate absorber tube enhanced the temperature profile and decreased the optimum deflection by 45–49% compared to steel. Hence, this approach could be used as a quick and effective approach to obtain the structural performance under non-uniform thermal conditions.

Okonkwo et al. (2018) modelled a commercially available PTC using the engineering equation solver (EES). The study compared the exergetic performance (exergy losses and exergy destruction) of four different geometries: conventional tube, longitudinal finned tube, tube with twisted tape insert and converging-diverging tube as shown in Fig. 3.79. Different combinations of inlet temperature and volumetric flow rate were studied for Therminol VP-1 and Al_2O_3-oil nanofluids. The outcomes have shown that the converging-diverging geometry was observed to achieve the thermal and exergetic efficiencies of 65.95% and 38.24%, respectively. It was found that the optical losses accounted for the main cause of exergetic losses with 24.5% for all the examined cases. The exergy destroyed accounted for 59.7% in the

a   **b**

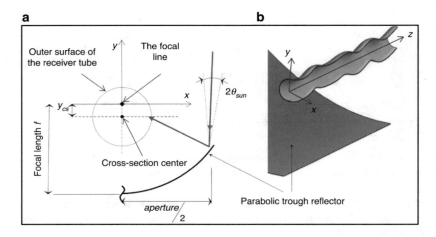

c

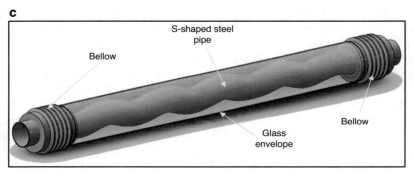

Fig. 3.75 (**a**) The representative cross-section, (**b**) setup of the novel PTR and (**c**) 3D schematic view of the PTR Bitam et al. (2018) (License Number: 4615170673264)

converging-diverging tube at 350 K and 54.7% for the smooth absorber tube at the same temperature. It was inferred that the use of Al$_2$O$_3$-oil nanofluid with the converging diverging receiver enhanced the exergetic efficiency of the PTC by 0.73%.

Rawani et al. (2018a, b) used an oblique delta-winglet twisted tape insert (ODWTTI) in the PTC absorber tube and analysed its thermal performance analytically as shown in Fig. 3.80. The entropy generation, thermal and exergy efficiencies, fluid temperature rise and operating parameters influence on the thermal performance were examined by using heat transfer mathematical model developed with the aid of C^{++} language. Various twist ratios (x) in the range of 1–4 and Reynolds number in the range of 3000–9000 were used. The outcomes revealed that a higher thermal effectiveness of PTC was observed when using TTI. It was concluded that for x = 1, the Nusselt number was 3.24 times over PTC with plain absorber tube with the corresponding thermal efficiency enhancement of 12.05% and enhancement factor of. 1.121 for similar operations.

Rawani et al. (2018a, b) compared the performance of various types of twisted tape inserts, i.e., square cut, oblique delta-winglet, alternate clockwise and counterclockwise, and serrated in PTC absorber tube as shown in Fig. 3.81. They used their

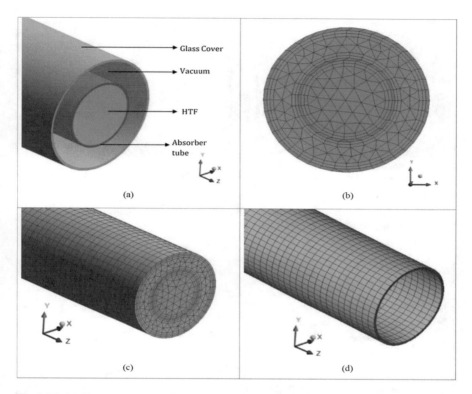

Fig. 3.76 (**a**) The computational domain, (**b**) Structure of grid in computational geometry (Front view), (**c**) Isometric view and (**d**) Structure of grid in the aborber tube Tripathy et al. (2018). (License Number: 4615170858690)

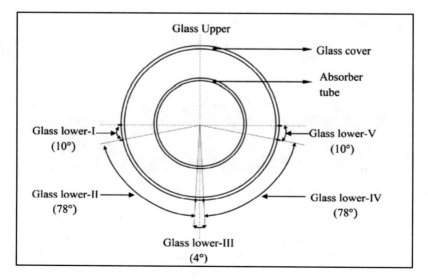

Fig. 3.77 Schematic diagram of the divison of glass cover outer surface Tripathy et al. (2018). (License Number: 4615170858690)

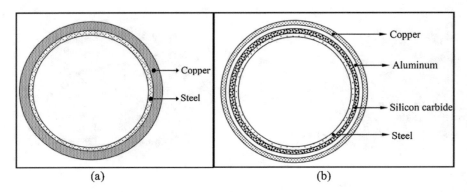

Fig. 3.78 (**a**) Bimetallic absorber tube and (**b**) Tetra-layered laminate absorber tube Tripathy et al. (2018). (License Number: 4615170858690)

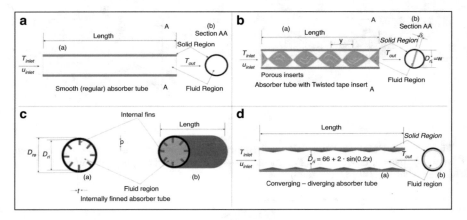

Fig. 3.79 Schematic illustration of (**a**) smooth tube, (**b**) tube with twisted tape insert, (**c**) internally finned tube and (**d**) converging-diverging tube Okonkwo et al. (2018). (License Number: 4615170858783)

Fig. 3.80 Oblique delta winglet twisted tape Rawani et al. (2018a, b)

developed mathematical model with the aid of C^{++} language to assess the PTC thermal performance. The results revealed that the serrated twisted tape inserts with $x = 2$ provides the highest performance compared with other inserts. It was found that the Nusselt number with serrated twisted tape insert ($x = 2$) is 3.56 times over PTC with plain absorber tube with thermal efficiency enhancement of 13.63% and

Sl. No	Types of twisted tape	Diagram
1	Plain twisted tape	
2	Square cut twisted tape	
3	Oblique delta-winglet twisted tapes	
4	Alternate clockwise and counter clock wise twisted-tape	
5	Serrated twisted tape	

Fig. 3.81 Various twisted tape inserts types used Rawani et al. (2018a, b)

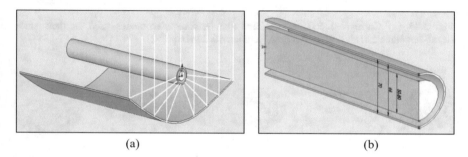

(a)   (b)

Fig. 3.82 (**a**) PTC system and (**b**) Absorber tube cross section Sadaghiyani et al. (2018)

exergy efficiency of 15.40%. It was concluded that the serrated twisted tape provides the lowest entropy generation.

Sadaghiyani et al. (2018) studied the influence of evacuated glass cover on the convective heat loss and exergetic efficiency. The evacuated receiver tube of LS-2 PTC was firstly simulated and analysed with CFD as shown in Fig. 3.82. Secondly, the collector and its absorber tube were simulated without evacuated glass cover for various wind speeds and collector orientations. Then, the exergetic analysis of each case-studies were calculated and the effect of wind blast and collector orientation on the exergy destruction and exergy loss were investigated. It was found that when wind blows on the convex side of the parabolic mirror, the impressibility of outlet temperature from wind speed is least than other orientations. Also, if the wind blows on the convex side of the parabolic mirror, the impressibility of outlet temperature from the variation of orientation is most than other orientations and consequently the exergy efficiency of collector was decreased. With increasing of wind blast, the exergy loss and the destruction of collector increased and consequently the exergy efficiency of collector was decreased. It was concluded that using evacuated tube leads to increasing of exergy from 10% to 60%.

Benabderrahmane et al. (2016) studied numerically the flow and thermal performance of tube receiver with central corrugated insert for PTC system as shown in Fig. 3.83. The Monte Carlo ray-tracing method (MCRT) coupled with finite volume method (FVM) was utilised. The outcomes revealed that the utilisation of corrugated insert could significantly augment the overall thermal effectiveness in the

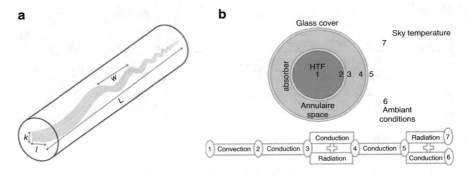

Fig. 3.83 (**a**) Corrugated PTC system and (**b**) heat tarnsfer modes used in their study Benabderrahmane et al. (2016). (License Number: 4031141658252)

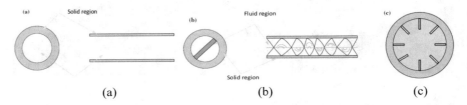

Fig. 3.84 (**a**) Smooth PTC tube, (**b**) PTC tube equipped with twisted tape insert and (**c**) PTC tube equipped with internal fins Khan et al. (2020). (License Number: 5030041356315)

range of 1.3–2.6. The increment of corrugation's twist ratio and the decrement of pitch between two corrugations increased the thermal performances.

Khan et al. (2020) compared numerically energetic and exergetic performance of absorber tube with twisted tape insert and tube with longitudinal fins using Al_2O_3/water with the aid of EES as shown in Fig. 3.84. The outcomes revealed that the absorber tube with twisted tape insert has the highest thermal efficiency of 72.26%, compared to a tube with internal fins (72.10%), and smooth absorber tube (71.09%). It was confirmed that the utilisation of nanofluid and passive techniques produced better thermal effectiveness.

A summary of literature studies on PTC utilising various passive techniques are listed in Table 3.2 chronologically.

It can be clearly noticed from Table 3.2 that the vast majority of the research efforts, which have been carried out by different researchers at different countries around the globe using various passive techniques with different shapes and configurations, are mainly done by China with 27%, followed by India with 20%, Greece with 10%, South Africa with 7% and other countries with 1–3% as can be seen in Fig. 3.85.

Table 3.2 Literature studies carried out to date on PTC using various passive techniques

References	Study type	Passive technique	Fluid used	Phase mode	Flow type	Remarks	Proposed correlations
Grald and Kuehn (1989)	Numerical/1D	Porous absorber	Different fluids of different thermal conductivities 0.01–0.8 W/m.°C	Single phase	Flow velocity was 8.11 × 10⁻⁵ m/s	Numerical results for thermal condctivity show the potentail for marked improvent over available parabloic trough collectors for a wide range of fluid outlet tempratures	Nil
Reddy et al. (2008)	Numerical	Porous finned receiver	Therminol-VP 1	Single phase	Turbulent flow Re number was in the range of 150,000–11,500,000	It was found that the porous inserts in PTC receiver improved the heat transfer by 17.5% with a pressure loss of 2 kPa	$Nu = 0.0885 \times Re^{0.75} (1 + A_s)^{-0.1427}$
Reddy and Satyanarayana (2008)	Numerical	Porous finned receiver	Therminol-VP 1	Single phase	Mass flow rate was in the range of 0.5–6.5 kg/s	The natural convection heat loss was lower for porous receiver than the plain receiver. The optimal thermal and flow fields were found with a trapezoidal porous fin of 4 mm thickness, tip-to-base thickness ratio (λ) of 0.25 and a ratio of consecutive fin distance to diameter of the receiver (l/d) of 1	Nil

(continued)

Table 3.2 (continued)

References	Study type	Passive technique	Fluid used	Phase mode	Flow type	Remarks	Proposed correlations
Kumar and Reddy (2009)	Numerical	Porous disc receiver	Therminol-VP 1	Single phase	Turbulent flow Re number was in the range of 40,000–250,000	The maximum heat transfer coefficient is achieved in top half-porous disc receiver with $H = 0.5di$, $w = di$ at $\theta = 30°$ with reasonable drag. The increase in Nusselt number for optimum receiver configuration is 64.2% compared to tubular receive with pressure drop of 457 Pa	Nil
Munoz and Abanades (2011)	Numerical/ CFD	Helical internal fins	Syltherm 800 oil	Single phase	Mass flow rate was 47.7 L/min	The results provide a collector enhancement efficiency of 3% and increase of pressure losses of 40%. This would lead to a 2% plant performance enhancement with a reduction of operation and maintenance costs	Nil

Kumar and Reddy (2012)	Numerical	Porous finned receiver with different inclinations	Therminol-oil 500 and water	Single phase	Mass flow rate was in the range of 0.5–1 kg/s	The optimum porous disc receiver improved the heat transfer rate and reduced the pumping power losses using therminol oil-55 higher than water as compared to plain receiver

$$Nu_w = 0.18 \left(\frac{\varnothing}{\varnothing_{max}}\right)^{-0.085} \left(\frac{\theta}{\theta_{max}}\right)^{-0.172} \left(\frac{W}{d_i}\right)^{-0.183} \left(\frac{H}{d_i}\right)^{-0.263} \left(\frac{t}{d_i}\right)^{0.122} Re^{0.552} Pr^{0.863}$$

$$f_w = 0.694 \left(\frac{\varnothing}{\varnothing_{max}}\right)^{-0.847} \left(\frac{\theta}{\theta_{max}}\right)^{0.194} \left(\frac{W}{d_i}\right)^{-0.27} \left(\frac{H}{d_i}\right)^{2.693} \left(\frac{t}{d_i}\right)^{-0.054} Re^{-0.239} Pr^{1.380}$$

$$Nu_T = 2.917 \left(\frac{\varnothing}{\varnothing_{max}}\right)^{-0.15} \left(\frac{\theta}{\theta_{max}}\right)^{-0.157} \left(\frac{W}{d_i}\right)^{-0.424} \left(\frac{H}{d_i}\right)^{0.896} \left(\frac{t}{d_i}\right)^{0.228} Re^{0.509} Pr^{0.345}$$

$$f_T = 2228.33 \left(\frac{\varnothing}{\varnothing_{max}}\right)^{-0.107} \left(\frac{\theta}{\theta_{max}}\right)^{1.266} \left(\frac{W}{d_i}\right)^{-0.479}$$

(continued)

Table 3.2 (continued)

References	Study type	Passive technique	Fluid used	Phase mode	Flow type	Remarks	Proposed correlations
Cheng et al. (2012a, b)	Numerical	Unilateral longitudinal vortex generators	Oil syltherm-800	Single phase	Turbulent flow Re number was 3.8×10^4– 7.0×10^5	The thermal loss of the PTR reduced by 1.35–12.1% to that of the smooth PTR. The comprehensive enhanced heat transfer performance is higher when the re number is larger at the same inlet temperature	Nil
Islam et al. (2012)	Numerical	Flow restriction device (plug)	Syltherm-800	Single phase	Turbulent flow Re number was in the range of 14,235–15,943	They have identified the non-uniformity of the heat flux and temperature profiles around the tube were physical issues. The results affirmed that the optical simulation and computational heat transfer simulation can be done separately	Nil
Aldali et al. (2013)	Numerical	Helical internal fins	Water	Single phase	Velocity was in the range of 0.257– 0.354 m/s	The thermal gradient for the pipe without a helical fin is significantly higher compared with the pipe with helical fins. Aluminium pipe has much lower thermal gradient compared to steel pipe	Nil

| Wang et al. (2013) | Numerical | Metal porous foam | Superheated steam | Single phase | Re number was in the range of 1×10^5–1×10^6 | The highest thermal performance is achieved when geometrical parameter (H) equals to 0.75. The Nu number increased 10–12 times associated with friction factor increment of 400–700 times and the PEC increased from 1.1 to 1.5. The thermal stresses were significantly reduced when the circumfere-ntial temperature difference on the receiver's tube outer surface decreased by 45% | Nil |
| Ghasemi et al. (2013) | Numerical | Two segmental rings | Therminol 66 | Single phase | Re number was in the range of 30,000–250,000 | The inclusion of porous rings enhanced the thermal efficiency and the increases of distance between two rings reduced the Nusselt number. The optimum thermal performance was achieved for the case when "Q" = 0.75d "the distance between porous two segmental rings" and "d/D" = 0.6 "inner diameter/inner diameter ratio", respectively | Nil |

(continued)

Table 3.2 (continued)

References	Study type	Passive technique	Fluid used	Phase mode	Flow type	Remarks	Proposed correlations
Ghadirijafarbeigloo et al. (2014)	Numerical	Perforated louvered twisted tape (PLTT)	Behran thermal oil	Single phase	Turbulent flow Re number was in the range of 5000–25,000	PLTT provided higher Nusselt number and friction factor compared to a smooth tube. It is observed that LLT can provide a maximum of 150% enhancement for Nusselt number and 210% for friction factor	Nil
Too and Benito (2013)	Mathematical	Helical coil / wire insert, twisted tape insert, dimpled tube, porous Foam	Air, CO_2, and Helium	Single phase	Turbulent flow Re number was in the range of 7083–42,499	Unrivaled overall thermal and hydraulic performance was obtained when using tubes with deeper protrusions compared to a plain tube. Working gases such as CO_2 and He can decrease the total pressure drop in a solar tubular gas receiver	Nil
Waghole et al. (2014)	Experimental	Twisted tape inserts	Silver-water Nanofluid	Single phase	Turbulent flow Re number was in the range of 500–6000	The thermal and hydraulic performance of silver nanofluid is higher compared to water in a PTC receiver. The Nusselt number, friction factor and enhancement efficiency were observed to be 1.25–2.10 times, 1.0–1.75 times and 135–205%, respectively, compared with the plain receiver of PTC	For water with twisted tape inserts: $Nu = 1.2546\,Re^{0.2843}Pr^{0.4}($ $1 + H/D)^{0.004}$ $f = 0.0841\,Re^{-0.2729}\,(1 + H/D)^{-0.1701}$ For silver nanofluid with twisted tape inserts: $Nu = 1.3167\,Re^{0.2843}Pr^{0.4}($ $1 + H/D)^{0.004}Pr^{0.4}(1 + \varnothing)^{-0.008}$ $f = 0.0841\,Re^{-0.2729}$ $(1 + H/D)^{-0.1701}(1 + \varnothing)^{-0.5}$ $(Nu_w/Nu) = 1.4058 - (0.0280/y)$

Song et al. (2014)	Numerical	Helical screw-tape inserts	Downtherm-A	Single phase	Turbulent flow Re number was in the range of 5000 –75,000	The influence of the transversal angle (β) on the heat flux distribution was found more significant than the longitudinal angle (φ). It is observed that the helical screw-tape inserts remarkably decreased the heat losses and maximum and circumferential temperature differences	Nil
Mwesigye et al. (2014a, b)	Numerical	Perforated plate insert	Sytherm-800	Single phase	Turbulent flow Re number was in the range of 1.02×10^4– 7.38×10^5	The inclusion of perforated plate inserts was shown to enhance the thermodynamic performance and to reduce the temperature gradients of the receiver. The modified thermal efficiency increased between 1.2% and 8%	$$Nu = \frac{5.817\, Re^{0.9483}\, Pr^{0.4050}\, \varsigma_p^{-0.1442}\, \varsigma_d^{0.4568}}{1000}$$ $(1+0.0742\tan\beta)$ $$f = \frac{0.1713\, Re^{-0.0267}\, \varsigma_p^{-0.8072}\, \varsigma_d^{3.1783}}{(1+0.08996\sin\beta)}$$
Syed Jafar and Sivaraman (2014)	Experimental	Nail twisted tapes inserts	Al_2O_3 - water nanofluid	Single phase	Laminar flow Re number was in the range of 710–2130	The insertion of nail twisted tape in an absorber of PTC using nanofluids can greatly enhance its thermal performance and increase its friction factor	Nil

(continued)

Table 3.2 (continued)

References	Study type	Passive technique	Fluid used	Phase mode	Flow type	Remarks	Proposed correlations
Sadaghiyani et al. (2014)	Numerical	Plug tube insert	Sytherm-800, Downtherm-rp, downtherm-j	Single phase	Mass flow rate was 0.6782 kg/s	The results show that the natural convection overcome on the forced convection, for r* < 0.6 m, and the mixed convection is the dominant mechanism of heat transfer for r* > 0.6 m. the friction factor decreased as the plug diameter increased and Nusselt number was minimum at r* = 0.6 m	Nil
Mwesigye et al. (2015)	Numerical	Perforated plate insert	Al_2O_3– synthetic oil	Single phase	Turbulent flow Re number was in the range of 3560–1.15×10^6	The thermal performance and the thermal efficiency increased up to 38% and 15%, respectively. It was found that high inlet temperatures and low flow rates show remarkable enhancement in the receiver's thermal efficiency	Nil
Mwesigye et al. (2015)	Numerical/ optimisation	Perforated plate insert	Sytherm-800	Single phase	Turbulent flow Re number was in the range of 1.02×10^4–1.36×10^6	It was found that the entropy generation decreased when the orientation angle increased. Increasing the plate size and decreasing the plate spacing decreased the optimal re number. The entropy generation rates are reduced by 53% below the optimal re number	Nil

Huang et al. (2015)	Numerical	Helical fins, protrusions and dimples	Therminol-VP1	Single phase	Turbulent flow Re number was in the range of $1 \times 10^4 – 2 \times 10^4$	The dimpled receiver tubes have unrivalled thermal performance compared with that having helical fins or protrusions. The dimples with deeper depth, narrower pitch and more numbers in the circumferential direction enhanced the thermal performance while other arrangements shown insignificant effect	Nil
Chang et al. (2015)	Numerical	Twisted tape inserts	Molten salt	Single phase	Turbulent flow Re number was in the range of 7485–30,553	Insert twisted tape significantly improved the uniformity of temperature distribution of tube wall using molten salt. The decreases of clearance rate © and twisted rate enhanced the heat transfer effectively. The decreases of clearance rate and twisted rate lead to increase the friction factor	Nil

(continued)

Table 3.2 (continued)

References	Study type	Passive technique	Fluid used	Phase mode	Flow type	Remarks	Proposed correlations
Lu et al. (2015)	Theoretical	Spirally grooved pipe	Molten salt	Single phase	Turbulent flow Re number was in the range of 5000–15,000	The increment of flow velocity and groove height increased the absorption efficiency and decreased the wall temperature. The heat absorption efficiency increased by 0.7%, and the maximum bulk fluid temperature increased to 31.1 °C	Nil
Zhu et al. (2015)	Experimental	Coiled receiver	Air	Single phase	Turbulent flow Mass flow rate was 0.045 kg/s	The highest energy efficiency of the solar receiver is about 82%, while it is maintained at about 80 for steady state and the highest exergy efficiency is about 28%	Nil
Reddy et al. (2015)	Experimental work	Porous disc enhanced receiver	Water and oil	Single phase	Turbulent flow Flow rate was in the range of 100–1000 L/h	The temperature gradient between the receiver's wall surface and fluid was less when using porous disc enhanced receiver compared to classical tubular receiver. The performance of PTC with porous disc enhanced receiver is much better than other receiver configurations	Nil

Diwan and Soni (2015)	Numerical	Wire-coils insert	Water	Single phase	Turbulent flow Mass flow rate was in the range of 0.01388–0.099 kg/s	The introduction of wire coils increased the Nusselt number by 104% to 330%. Good thermal performance was obtained when using wire-coils inserts with pitch value from 6 to 8 mm at lower flow rates and using pitch of 8 mm at higher flow rate	Nil
Mwesigye et al. (2016a, b)	Numerical	Wall-detached twisted tape inserts	Sytherm-800	Single phase	Turbulent flow Re number was in the range of 10,260–1,353,000	The increase of thermal performance of 169% and increase in thermal efficiency up to 10% were achieved compared with a plain absorber tube. A 58% decrease in the entropy generation rate was also attained	$Nu = 0.01709\,Re_p^{\,0.8933}\,Pr^{0.3890}\,y^{-0.4802}\,\varsigma_w^{0.3881}$ $f = 1.1289\,y^{-1.0917}\,\varsigma_w^{1.1802}$
Nems and Kasperski (2016)	Experimental	Internal multiple-fin array	Air	Single phase	Flow rate range was in the range of 0.001–0.009 m^3/s	The prototype design suffered from two weak sides: Lack of tightness of the absorber and lack of vacuum control of the glass casing. They suggested that the absorber should be longitudinally welded, covered with selective oxide layer and mounted inside a single glass pipe. It enables to obtain vacuum from the glass cover, which will allow to increase thermal efficiency	Nil

(continued)

Table 3.2 (continued)

References	Study type	Passive technique	Fluid used	Phase mode	Flow type	Remarks	Proposed correlations
Bellos et al. (2016a, b)	Numerical	Converging–diverging absorber tube	Thermal oil, thermal oil-nanoparticles (Al_2O_3), pressurised water	Single phase	Turbulent flow Re number was in the range of 4000–25,000	The utilisation of Al_2O_3-thermal oil nanofluid improved the mean efficiency by 4.25% while the use of pressurised water showed 4.55% mean efficiency enhancement compared to plain tube geometry. At higher fluid temperature levels, the increase in the efficiency was greater	Nil
ZhangJing et al. (2016)	Numerical (optimised by genetic algorithm (GA))	Porous insert	Water/steam	Single phase	Turbulent flow Re number was in the range of 50,000–900,000	The porous insert receiver tube exhibited higher thermal performance than that of the plain receiver. It achieved flawless thermo-hydraulic performance by using similar optimised porous insert, which is difficult to achieve by utilising the plain receiver	Nil
Fuqiang et al. (2016a, b)	Numerical	Asymmetric outward convex corrugated tube	Thermal oil- D12	Single phase	Turbulent flow Re number was in the range of 10×10^3 to 98×10^3	The utilisation of asymmetric outward convex corrugated tube as a receiver of PTC improved the thermal performance and decreased the thermal strain effectively. A maximum 148% and 26.8% of overall thermal performance and von-Mises thermal strain, respectively, were achieved	$Nu_A = 2.665 \times 10^{17} \left(\frac{p}{D}\right)^{-0.135} \left(\frac{H}{D}\right)^{0.027} \left(\frac{rl}{D}\right)^{-0.008}$ $Re^{1.682} Pr^{-24.243}$ $f_A = 0.203 \left(\frac{p}{D}\right)^{-0.262} \left(\frac{H}{D}\right)^{0.1} \left(\frac{rl}{D}\right)^{-0.042} Re^{-0.246}$

Reference	Method	Geometry/Insert	Fluid	Phase	Flow conditions	Findings	Correlation
Jaramillo et al. (2016)	Numerical	Twisted tape insert	Water, air	Single phase	Turbulent flow Re number was in the range of 1389.7–8338.03	The Nusselt number, the removal factor, the friction factor and the thermal efficiency increased as both the twist ratio (y/w) and the Reynolds number decreased. These quantities did not present an enhancement when the twist ratio increased	Nil
Fuqiang et al. (2016a, b)	Numerical/ CFD and FEM	A symmetric outward convex corrugated tube	Thermal oil- D12	Single phase	Re number was in the range of 16,000–80,000	The utilisation of symmetric outward convex corrugated tube provided an increment of thermal performance up to 8.4% and maximum decrement of the thermal strain of metal tube up to 13.1%	$Nu_{eff} = 0.8114\ Re^{0.711} Pr^{0.379}$ $Nu_{eff} = \left(8.157 \times 10^{8}\right) Re^{0.464} Pr^{4.011} \left(\dfrac{p}{D}\right)^{-0.041}$
Benabderrahmane et al. (2016)	Numerical 3D	Longitudinal rectangular/ triangular fins	Downtherm-A oil with different nanofluids Al_2O_3, cu, sic, C and cu	Single phase	Turbulent flow Re number was in the range of 2.57×10^{4}– 2.57×10^{5}	It is observed there was a remarkable heat transfer enhancement from 1.3 to 1.8 times. The cu and SiC nanoparticles improved heat transfer significantly than C and Al_2O_3 nanoparticles. The utilisation of fins and nanofluid provided greater thermo-hydraulic performance	Nil

(continued)

Table 3.2 (continued)

References	Study type	Passive technique	Fluid used	Phase mode	Flow type	Remarks	Proposed correlations
Chang et al. (2017)	Numerical	Concentric and eccentric pipes	Molten salt	Single phase	Turbulent flow Re number was in the range of $1 \times 10^4 – 9 \times 10^4$	Inserts can significantly improve heat transfer performance of more than 1.64 times than a PTR without inserts when the PTR is inserted by A3 (larger diameter). The eccentric pipe inserts of B3 (larger eccentricity) performed significantly better than the concentric tube inserts for its excellent performance in decreasing the maximum temperature of absorber tube and molten salt	Nil
Ghasemi and Ranjbar (2017a, b)	Numerical	Porous rings	Syltherm 800	Single phase	Turbulent flow Re number was in the range of 30,000–25,0000	The heat transfer characteristics enhanced by inserting the porous rings in tubular solar absorber. The decrement in the distance between porous rings and the increment of the inner diameter of the porous rings would increase and decrease the heat transfer, respectively	Nil

Bortolato et al. (2017)	Experimental	Flat absorber	SWCNH – water nanofluid	Single phase	Mass flow rate was 350 kg/h	The application of a carbon nanohorn-based nanofluid displayed an efficiency comparable to that obtained with a surface receiver tested in the same system. However, such performance was not maintained for a long time because of lack of stability of the absorbing fluid	Nil
Rawani et al. (2017)	Mathematical	Serrated twisted tape insert	Therminol VP-1	Single phase	Turbulent flow Re number was in the range of 3000–9000	The thermal efficiency enhancement was 15.7%, while the exergy efficiency was 12.10% and enhancement factor was 1.157 for similar conditions	Nil
Jamal-Abad et al. (2017)	Experimental	Copper porous media	Water	Single phase	Flow rate was in the range of 0.5–1.5 L/min	An improvement in Nu number was observed by using metal foam. The friction factor increased considerably when tube was filled with metal foam. The removed energy and absorbed energy parameter of the collector decreased by applying copper foam inside the absorber	Without copper foam, $U_L = 28.33$, $F_R = 0.88$ $\eta = -2.256\,x + 0.5547$ With copper foam, $U_L = 15.44$, $F_R = 0.854$ $y = -1.1934\,x + 0.5381$

(continued)

Table 3.2 (continued)

References	Study type	Passive technique	Fluid used	Phase mode	Flow type	Remarks	Proposed correlations
Pavlovic et al. (2017a, b)	Experimental and numerical	Corrugated spiral cavity receiver	Water, air and Therminol VP-1	Single phase	Turbulent flow mass flowrate was in the range of 100–350 L/h	Water and the thermal oil were the most selected working fluids compared with other fluids at low and high temperatures, respectively. Air and thermal oil exergetically were the best selection in low and high temperatures, respectively	Nil
Bellos et al. (2017a, b, c)	Numerical	Internally finned/ evacuated tube absorber	Oil syltherm-800	Single phase	Turbulent flow mass flowrate was in the range of 50–250 L/min	The optimum case was achieved with an absorber with 10 mm fin length and 2 mm fin thickness. The thermal efficiency was improved around 0.82%, the Nusselt number increased 65.8%, whilst the friction factor and the pressure losses were doubled compared to the smooth case	$Nu = 0.01638\,Re^{0.851}$ $Pr^{0.374}\left[1+0.4442\left(\dfrac{t}{D_{ri}}\right)^{0.270}\left(\dfrac{p}{D_{ri}}\right)^{1.024}\right]$ $f = \dfrac{0.2585}{Re^{0.2386}}\left[\dfrac{1+2.7452\left(\dfrac{t}{D_{ri}}\right)^{0.118}\left(\dfrac{p}{D_{ri}}\right)^{0.839}}{\exp\left(9.711\dfrac{t}{D_{ri}}\right)\exp\left(4.01\dfrac{p}{D_{ri}}\right)}\right]$
Bellos et al. (2017a, b, c)	Theoretical and numerical	Internally finned absorber	Oil syltherm-800	Single phase	Turbulent flow mass flowrate was in the range of 50–250 L/min	The optimum case with 20 mm fin length and 4 mm fin thickness was achieved. The thermal efficiency and thermal enhancement index were increased by 1.27% and 1.483, respectively, whilst the Nusselt number was confirmed to be 2.65 times higher than the smooth case	Nil

Bellos et al. (2017a, b, c, d, e)	Theoretical	Internal longitudinal fins absorber	Air, Helium and CO_2	Single phase	Turbulent flow mass flowrate was in the range of 50–250 L/min	Higher exergetic efficiency was obtained with fins having 10 mm length. Helium was the suitable working fluid exergetically. Helium was the optimum solution up to 290 °C and after this point, carbon dioxide performed better	Nil
Xiaowei Zhu et al. (2017)	Numerical	Wavy-tape insert	Sytherm-800	Single phase	Turbulent flow The mass flowrate was in the range of 240–720 L/min	The Nusselt number was improved by 261–310%, which benefits to the decreases of both PTR structure temperature and total heat loss. The heat loss was found to be reduced by 17.5–33.1%, depending on the HTF flow rate	Nil
Bellos et al. (2017a, b, c, d, e)	Numerical	Internal fins	CO_2	Single phase	Mass flow rate mass flowrate was in the range of 0.1–0.25 kg/s	The thermal performance increased when using higher fin length, while the optimum fin length was 10 mm with 45.95% exergetic efficiency based on the exergy analysis. The pressure losses were the best criteria for assessing solar collectors working with gases	Nil

(continued)

Table 3.2 (continued)

References	Study type	Passive technique	Fluid used	Phase mode	Flow type	Remarks	Proposed correlations
Xiangtao et al. (2017)	Numerical/CFD	Pin fin array absorber	Thermal oil- D12	Single phase	Re number was in the range of 2000 to 18,000	The utilisation of tube receiver with pin fin arrays inserts increased the Nusselt number up to 9% and the thermal performance factor up to 12%	$\Delta T_{max-min} = 4495.99$ $-4210.30\left(1-\exp\left(-\dfrac{Re}{661.65}\right)\right)$ $-207.03\left(1-\exp\left(-\dfrac{Re}{13168.99}\right)\right)$ $TKE = 0.00143\exp\left(\dfrac{Re}{22304.39}\right) - 0.00156$
Huang et al. (2017)	Numerical	Dimpled receiver	Therminol-VP1	Single phase	Turbulent flow Re number was 2×10^4	The average Nusselt number and friction factor in dimpled receiver tubes under non-uniform heat flux (NUHF) were larger than those under uniform heat flux (UHF). The deep dimples ($d/D_t = 0.875$) were good performing compared with shallow dimples ($d/D_t = 0.125$) at similar Grashof number	Nil
Chang et al. (2018)	Numerical	Concentric rod and eccentric rod	Molten salt	Single phase	Turbulent flow Re number was in the range of 10,000–30,000	The temperature profile can be stabilised, and the maximum temperature can be significantly decreased with the increasing of dimensionless diameter (B) and dimensionless eccentricity (H), which helps to reduce the PTR thermal deflection and increase its reliability	Nil

Bellos et al. (2018a, b, c)	Numerical	Insert with star shape	Syltherm 800 (thermal oil)	Single phase	Volumetric flow rate was 150 L/min	The thermal efficiency enhancement increased when the inlet temperature increases. Higher dimensions of the insert provided to higher thermal performance enhancement. The use of inserts increased the pressure drop many times while the pumping loss was low in all cases studied	Nil
Bellos et al. (2018a, b, c)	Numerical	Fins with different locations and number	Syltherm 800 (thermal oil)	Single phase	Volumetric flow rate was 150 L/min	The optimum absorber was the one with three fins located in its lower part ($\beta = 0°$, $\beta = 45°$ and $\beta = 315°$) with thermal efficiency of 68.59%	Nil
Bellos et al. (2018a, b, c)	Numerical	Internal rectangular finned tube	CuO in Syltherm 800	Single phase	Volumetric flow rate was 150 L/min	The utilisation of internal fins and nanofluids exhibit an increment of 1.1% and 0.76% thermal efficiency enhancement, and the combination of both techniques provided 1.54%	Nil

(continued)

Table 3.2 (continued)

References	Study type	Passive technique	Fluid used	Phase mode	Flow type	Remarks	Proposed correlations
Bitam et al. (2018)	Numerical/CFD	S-curved sinusoidal tube receiver	Synthetic oil	Single phase	Mass flow rate was in the range of 2–9.5 kg/s	The Nusselt number and friction factor increased by 45–63%, less than 40.8%, respectively, which provided 135% maximum performance evaluation criteria. The maximum of 35 K decrement in the circumferential temperature difference resulted in the reduction of thermal stresses and heat losses	Nil
Kumar et al. (2018)	Experimental	Fins	TiO₂ – water nanofluid	Single phase	Not specified	The outlet temperature increased with increases in nanofluid concentration. Fins were utilised to enlarge the surface area so that heat transfer rate was also increased	Nil
Okonkwo et al. (2018)	Numerical	Longitudinal finned absorber, twisted tape tube, converging-diverging absorbers	Al₂O₃-Therminol VP-1 nanofluid	Single phase	Mass flow rate was in the range of 25–200 L/min	The converging-diverging absorber produced the best exergetic enhancement of 0.65% using Therminol VP-1 and 0.73% using Al₂O₃/Therminol VP-1 nanofluid	Nil

| Rawani et al. (2018a, b) | Mathematical | Oblique delta-winglet twisted tape insert | Therminol VP-1 | Single phase | Turbulent flow Re number was in the range of 3000–9000 | The Nusselt number increased by 3.24 times compared with a plain PTC. The corresponding thermal efficiency and exergy efficiency were 12.05% and 4.92% and the enhancement factor was 1.121 for similar conditions | Nil |
| Rawani et al. (2018a, b) | Mathematical | Square cut, oblique delta-winglet, alternate clockwise and counter-clockwise, and serrated twisted tape insert | Therminol VP-1 | Single phase | Mass flow rate was in the range of 0.06–0.16 kg/s | The serrated twisted tape inserts (x = 2) produced Nusselt number of 3.56 and 3.19 times over PTC plain absorber. The thermal efficiency enhancement was 13.63% and the exergy efficiency was 15.40%. The serrated twisted tape provided the lowest entropy generation | |

(continued)

Table 3.2 (continued)

References	Study type	Passive technique	Fluid used	Phase mode	Flow type	Remarks	Proposed correlations
Sadaghiyani et al. (2018)	Numerical	Plug tube insert	Sytherm-800	Single phase	Mass flow rate was 0.6782 kg/s	It was observed that when wind blows on the convex side of the parabolic mirror, the impressibility of outlet temperature from wind speed is least than other orientations and this decreased the exergy efficiency. The exergy loss and the destruction of collector increased with increasing the wind blast. The use of evacuated tube led to increase the exergy efficiency from 10 to 60%	Nil
Benabderrahmane et al. (2016)	Numerical	Corrugated tube insert	Nitrate salt	Single phase	Turbulent flow Re number was in the range of 10^4–10^6	The outcomes revealed that the utilisation of corrugated insert could significantly augment the overall thermal effectiveness in the range of 1.3–2.6. The increment of corrugation's twist ratio and the decrement of pitch between two corrugations increased the thermal performances	$E = 8.68 \dfrac{T_{in} - T_{amb}}{G_b} + 0.79$
Khan et al. (2020)	Numerical	Absorber tube with twisted tape insert and tube with longitudinal fins	Al_2O_3/water	Single phase	Turbulent flow Re number was in the range of 10^3–$5*10^5$	The outcomes revealed that the absorber tube with twisted tape insert has the highest thermal efficiency of 72.26%, compared to a tube with internal fins (72.10%), and smooth absorber tube (71.09%)	Nil

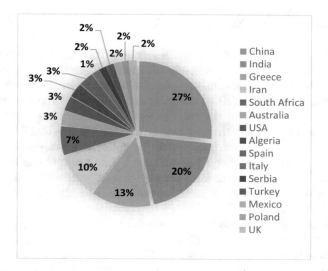

Fig. 3.85 Research efforts on PTC equipped with passive techniques around the globe

3.2.2 WPTC Studies Using Conventional Working Fluids

3.2.2.1 Introduction

Recently, PTC technology has been widely applied in concentrating power plants, hydrogen production or other industrial processes with high temperature requirement. In any case, the huge temperature gradient is the basic explanation of instigating the thermal twisting and harm of PTC. Accordingly, numerous scientists have embraced the strategy for heat transfer upgrade to diminish the temperature gradient by utilising diverse operating liquids. The regular working liquids utilised in PTC frameworks, e.g., water, thermal oils and molten salts cannot work in high temperatures. However, water/steam power plants depend on the requirements of high operating temperatures and the sophisticated control strategies. The utilisation of thermal oils as Syltherm 800, Therminol VP-1, Therminol 66, Therminol D-12, Dowtherm A and Behran oil is common in indirect frameworks with heat exchangers for the heat creation. Interestingly, the thermal oils can work up to 400 °C with a sensible pressure level (close to 15 bars). However, the utilisation of thermal oils prompts moderately low thermal execution, and it requires high maintenance cost. The up-and-coming age of working liquid in PTC systems is molten salts (particularly nitrate salts) giving higher edge in the sun-oriented energy to power conversion. Molten salts are normally nitrate salts, e.g., 60% $NaNO_3$–40% KNO_3, which can work up to 550 °C, giving conceivable outcomes for higher thermal effectiveness. However, these working liquids required high security in the activity because of the freezing threat which can be happened in temperature levels from 100 to 230 °C. In this manner, the molten salts must be held upper to a lower limit near 200 °C because of solidification threat. Different types of fluids such as liquid metals as sodium and gases (air, helium, nitrogen and CO_2) at higher temperatures can also be utilised but their operation is not yet settled at this point.

3.2.2.2   Numerical and Experimental Studies

There are numerical and experimental research articles have been conducted by diverse investigators on thermal augmentation using different types of classical fluids in PTC, which are comprehensively summarised below, in the following sections.

Arasu and Sornakumar (2006) designed and developed experimentally a solar PTC of aperture area 1m^2 for hot water generation as shown in Fig. 3.86. The collector's experiments were performed based on ASHRAE Standard 93, 1986. It was found that the test slope and intercept of the collector efficiency equation were 0.3865 and 0.6905, respectively. The collector's time constant and the collector's half acceptance angle obtained from the experiments were 67 s and 0.5°, respectively, which assumes that the collector operates consistently at most extreme conceivable effectiveness.

Cheng et al. (2010) analysed numerically 3D coupled thermal characteristics of Syltherm 800 oil in a PTC receiver tube as shown in Fig. 3.87. The MCRT method and CFD simulation using a commercial FLUENT software were combined and used. Three typical testing models were chosen (no-wall model, no-radiation model and unbridged model) to provide an extra illustration of the thermal mechanism in PTC tube. The outcomes indicated that the radiation loss in Model 3 is up to 153.70 W/m^2. It was revealed that the radiation loss should be decreased as much as possible to enhance the PTC's efficiency. The proposed 3D numerical models and methods can not only be used effectively to study the coupled heat transfer characteristics on known factors but also those difficult or costly to be tested.

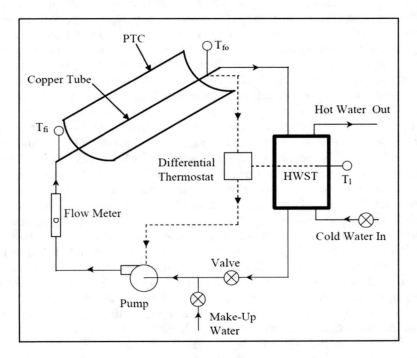

Fig. 3.86 PTC system Arasu and Sornakumar (2006)

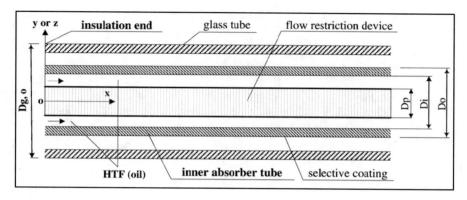

Fig. 3.87 Longitudinal cross-section of PTC receiver tube segment Cheng et al. (2010). (License Number: 4615211166760)

Tao and He (2010) developed a 2D numerical model for coupled heat transfer process in a PTC tube as shown in Fig. 3.88. The local temperature gradient, and local Nu number were investigated. The outcomes show that the natural convection influence must be considered when Ra number is larger than 10^5. When the tube diameter ratio increased, the Nusselt number in inner tube (Nu_1) increased and the Nusselt number in annuli space (Nu_2) decreased. It was found that Nu_1 decreased and Nu_2 increased when the tube wall thermal conductivity increased. However, the Nu number and average temperatures were less affected when thermal conductivity is greater than 200 W/m.K. It was observed that, because of the natural convection influence, the local Nu number on the inner tube increased until it achieved its optimum value then it gradually diminished again.

He et al. (2011) solved the complex coupled heat transfer problem in a PTC system using coupled simulation method of MCRT and FVM as shown in Fig. 3.89. The influences of different parameters such as proportions ratio (CR) and rim angles were investigated. The outcomes revealed that when CR increased, the heat flux profiles look smoother, the angle span became larger and the shadow influence of absorber tube got more vulnerable. However, when the rim angle raised, the heat flux maximum value became lower, and the curve moved towards the direction of $\varphi = 90°$. Nevertheless, the temperature increase was only improved when CR increased and when the rim angle greater than 15° its impact on the thermal fields was discarded. It was also noticed that when more rays were reflected by the glass tube, the temperature rise is much lower when the rim angle is around 15°.

Cheng et al. (2012a, b) combined FVM and MCRT method to solve 3D computational domain of PTC framework as illustrated in Fig. 3.90. Different working fluid types such as Therminol 1P1, Syltherm 800 and residual gas were used and examined. The results show that the properties of these fluids and operating conditions affected the thermal and flow fluid fields and the receiver's temperature profile, thus affected the thermal loss and the collector efficiency. It was concluded that the average Nusselt number of the two synthetic oil increased and the average friction factor decreased when the inlet fluid temperature increased. It was also observed that the two synthetic oil fluids have a better thermal effectiveness and a lower

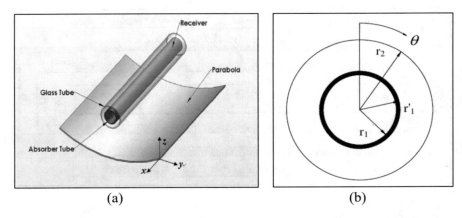

(a)   (b)

Fig. 3.88 (**a**) PTC system and (**b**) cross-section of the collector tube Tao and He (2010). (License Number: 4615750132592)

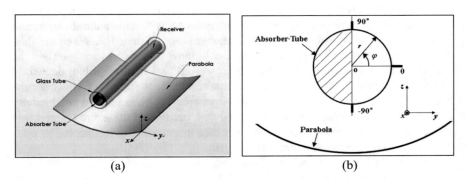

(a)   (b)

Fig. 3.89 (**a**) Schematic of a PTC and (**b**) The cylindrical coordinate He et al. (2011). (License Number: 4615750298931)

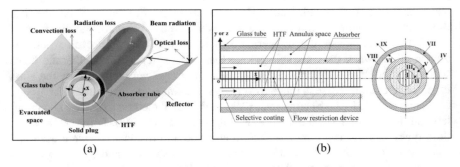

(a)   (b)

Fig. 3.90 (**a**) Physical model of a PTC and (**b**) Schematic of the photo-thermal computational model of a PTC Cheng et al. (2012a, b). (License Number: 4615750514622)

pressure drop than that of the two salts. The thermal loss of four residual gas conditions increased and the collector efficiency decreased when the fluids inlet temperature increased, respectively. The portions of the total thermal loss caused by radiation increased when fluids inlet temperature increased.

Hachicha et al. (2013) presented heat transfer model which includes different detailed elements of the receiver of PTC using finite volume method. The energy balance equation was implemented for each model's element along with the ray trace techniques as shown in Fig. 3.91. They used crossed string method to include the thermal radiative heat transfer. The results obtained were compared with the available experimental ones, with some observed variation at higher temperatures. These variations were caused by the optical properties of the heating element, and potential mistakes came from the utilised heat transfer coefficient relationships. It was thus inferred that the developed model was feasible for estimating the heating element optical and thermal patterns under various working situations.

Mwesigye et al. (2013) performed numerical simulation for entropy generation analysis in a PTC receiver at various parameter values as shown in Fig. 3.92. Using the first law of thermodynamics, the results revealed that the Nusselt number and friction factor did not change when the proportion ratios varied. The second law of thermodynamics indicated that the entropy generation values in the receiver's tube

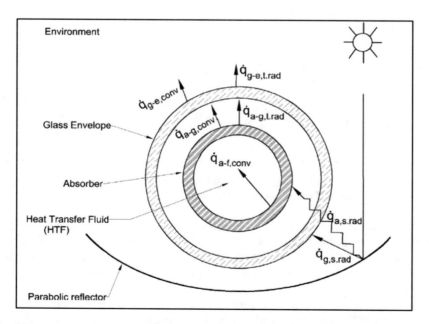

Fig. 3.91 (**a**) Schematic of the heat transfer model of PTC tube Hachicha et al. (2013). (License Number: 4615750743144). (**b**) (a) Longitudinal discretisation of the PTC tube, (b) Azimuthal discretisation of the PTC, (c) working fluid control volume, (d) Absorber tube control volume and (e) Glass tube control volume Hachicha et al. (2013). (License Number: 4615750743144)

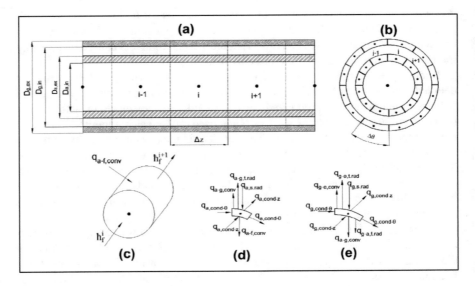

Fig. 3.91 (continued)

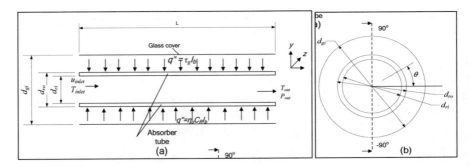

Fig. 3.92 (**a**) Physical model of a longitudinal PTC and (**b**) cross-section of computational domain Mwesigye et al. (2013). (License Number: 4615760049398)

increased as the proportion ratios increased. The results also show that at low flow rate values the Bejan number was around 1 and at the highest flow rate it was between 0 and 0.24. It was found that the entropy generation could be minimised when the entropy generation caused by the thermal and fluid irreversibilities in the receiver's tube is minimised.

Mwesigye et al. (2014a, b) investigated numerically the minimum entropy generation in a PTC receiver using MCRT method and CFD as shown in Figs. 3.93 and 3.94. The influences of rim angle, proportion ratio, Reynolds numbers and fluid temperature were examined. The results revealed that the entropy generation and the Bejan number both increased as the proportion ratio increased, and the rim angle decreased. The Bejan number expansion is an extra indication that the entropy

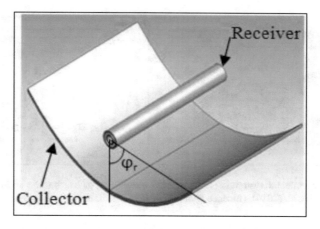

Fig. 3.93 3-D model of a PTC Mwesigye et al. (2014a, b). (License Number: 4615760387752)

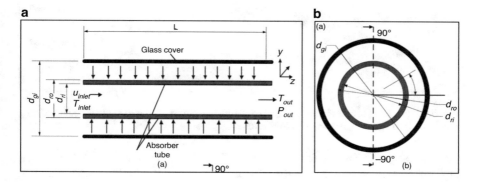

Fig. 3.94 (**a**) Physical model of a longitudinal PTC and (**b**) cross-section of computational domain Mwesigye et al. (2014a, b). (License Number: 4615760387752)

generation increment is caused by the thermal irreversibility as the proportion ratio expanded and the rime angle decreased.

Chang et al. (2014) presented experimental and numerical analysis of thermal effectiveness in PTC absorber pipes utilising non-uniform heat flux in high Reynolds number range as shown in Figs. 3.95 and 3.96. The results revealed that the fluid temperature and pipe wall profiles are very irregular in all three directions. The wall temperature profile of an absorber tube varied with the tubes' angle θ, and the maximum temperature happened at $\theta = 180°$. An empirical correlation for the wall temperature profile was proposed.

Wang et al. (2014) examined the thermal physics mechanisms of PTC systems using a 3D simulation based on FVM to solve the coupled problem as shown in Fig. 3.97. The performance of the PTCs using molten salt as a medium fluid was numerically studied, and the influences of other paramount parameters on the PTCs

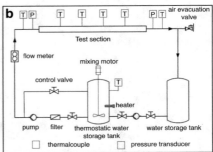

Fig. 3.95 Experimental system (**a**) experimental installation of test section and (**b**) Schematic diagram Chang et al. (2014). (License Number: 4615760389838)

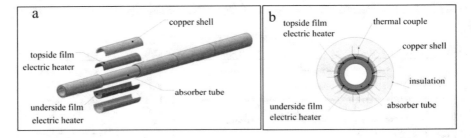

Fig. 3.96 (**a**) Test section diagram and (**b**) Schematic diagram of the tube cross-section Chang et al. (2014). (License Number: 4615760389838)

were investigated. The results revealed that the CTD of the absorber increased with DNI rising and decreased with the increase of medium fluid inlet temperature and inlet velocity. Furthermore, the numerical outcomes indicated that the non-uniform heat flux significantly affected the CTD of the absorber while it has a minimal impact on the thermal efficiency. The PTC's thermal efficiency using molten salt at 773 K is 7.90% lower than that using oil for typical work conditions.

Wu et al. (2014) simulated three-dimensional optics and temperature profile of a PTC receiver using the combination of a MCRT method and commercial software (FLUENT) as shown in Fig. 3.98. The coupled heat transfer mechanisms and fluid flow were together considered as shown in Fig. 3.99. The detailed temperature history of the whole PTC receiver was obtained. It was observed that the heat loss of the bellows was around 7% of the total heat loss if a heat convection coefficient 10 W/m² K was subjected on the bellows. The temperature difference within the metal tube changes inversely with HTF velocity. In addition, the simulation results show the stagnation temperature of PTC receiver increased linearly with time, and can reach at 700 K in 130 s.

Cheng et al. (2014) presented detailed comparative and sensitive analyses for different PTC systems to optimise their thermal and optical effectiveness as shown in Fig. 3.100. They used LAT73 Trough, T6R4 PTC system of different aperture

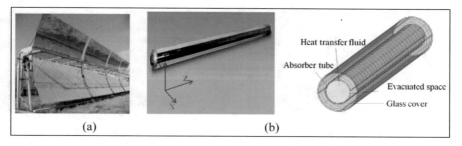

Fig. 3.97 (a) Experimental platform with 600 m² in Langfang city used for modelling and (b) Schematic diagram of the PTC Wang et al. (2014). (License Number: 4615770404855)

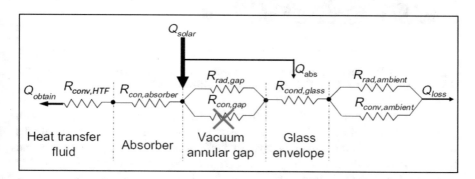

Fig. 3.98 (a) Structures of the studied PTC receiver and (b) cross-sectional of PTR with thermal resistances and heat flows Wu et al. (2014). (License Number: 4615770618019)

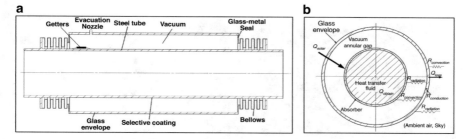

Fig. 3.99 Thermal network of parabolic receiver tube. (The red cross on $R_{con,gap}$ refers it is not considered in this study) Wu et al. (2014). (License Number: 4615770618019)

widths and T6R4 PTC system of different focal lengths under various working conditions. They utilised a previously developed unified MCRT model for the above-mentioned PTC systems along with various geometrical factors. The results revealed that the PTC frameworks had various levels of sensitivity to various optical errors, with consistent optical accuracy requirements. When the aperture width is larger there will be a high sensitivity of the PTC system to the optical errors. Thus, an ideal

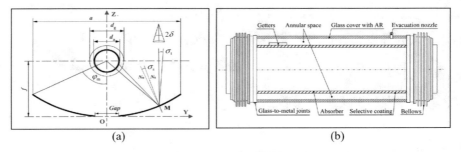

Fig. 3.100 (**a**) Schematic of cross-section of a PTC and (**b**) Schematic of the physical model of an evacuated tubular receiver Cheng et al. (2014). (License Number: 4615780487024)

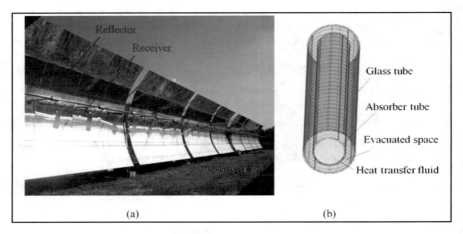

Fig. 3.101 (**a**) Schematic of experiment platform with size of 600 m² and (**b**) Schematic of a PTC solar receiver Wang et al. (2015). (License Number: 4615780671423)

optimised focal length of 0.47 m can be determined accordingly, with a maximum optical efficiency of 87.28%. The PTC systems of different active receiver lengths or different glass cover diameters have little effects on the optical performance and little sensitivity to the optical errors. However, the PTC systems of different absorber diameters have relatively larger effects than that of them. It was found that when the absorber diameter is smaller, there will be more sensitivity to the optical errors. However, there may be some contradictory trends in the optical and thermal efficiencies for optimising some geometric parameters, such as the glass cover diameter.

Wang et al. (2015) used a solar ray trace (SRT) and the finite element method (FEM) to analyse the coupled thermal and flow problem in a PTC framework as shown in Fig. 3.101. The heat flux trend was determined by the SRT method, and the influences of various working factors on the PTC's thermal effectiveness were numerically studied. The stress intensity and the thermal deformation profiles of the receiver were also examined. The results indicated that the CTD of the absorber

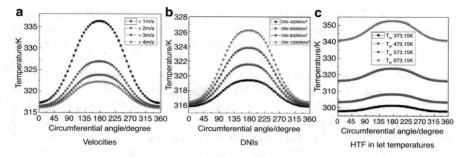

Fig. 3.102 Temperature distributions of the outer surface of the cover Wang et al. (2015). (License Number: 4615780671423)

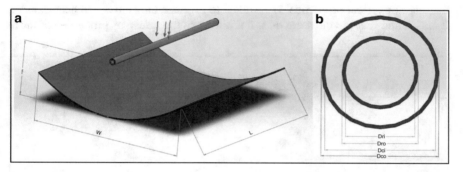

Fig. 3.103 PTC model designed in Solidworks, (**a**) trough dimensions and (**b**) evacuated tube dimensions Tzivanidis et al. (2015). (License Number: 4615760389921)

decreased with the increases of inlet temperature and velocity of the working fluid and increased with the increment of the DNI. It was found that the CTD of the absorber tube reached 22–94 K when the inlet velocity was in the range of 1–4 m/s, the DNIs were 500–1250 W/m^2 and the inlet temperature was in the range of 373–673 K as shown in Fig. 3.102. It was observed that the PTC's absorber tube thermal stress and deformations were greater than that of the glass tube.

Tzivanidis et al. (2015) designed and simulated a small PTC model for different operating conditions using commercial software Solidworks as shown in Fig. 3.103. The goal of this study was to predict the efficiency of this model and to analyse the heat transfer phenomena that takes place. The results show that the PTC's thermal efficiency is more than 75% for high temperatures. The temperature distribution is non-uniform in the peripheral of the absorber. It was concluded that the Reynolds number is fully reliable on the water inlet temperature. This illustrates that the flow is laminar at low temperature level with a convection coefficient about 300 W/m^{2}K and the flow becomes turbulent at higher temperatures with a greater convection coefficient about 1000 W/m^{2}K.

Li et al. (2015) analysed numerically mixed convective thermal effectiveness in a PTC receiver tube as shown in Fig. 3.104. The influence of buoyancy force under uniform heat flux (UHF) and non-uniform heat flux (NUHF) was analysed quantitatively. The effect of solar elevation angel on the thermal and flow fields was also investigated. The results revealed that there is noticeable difference in flow field and temperature profile between UHF and NUHF. The Nusselt number and friction factor of UHF case is greater than that under NUHF case for $\varphi = 0°$ and $30°$ and less than that under NUHF case for $\varphi = 60°$ and $90°$ as shown in Fig. 3.105. It was found that it is unfavourable to utilise the experimental correlations for combined/forced convection to carry out the thermal design of a PTC heated by NUHF.

Basbous et al. (2015) performed a detailed mathematical model to compare the PTC's thermal effectiveness installed in Casablanca of Morocco between humid and dry areas as shown in Fig. 3.106. The influence of the humidity of humid air on the thermophysical properties was studied to evaluate the convective heat transfer. It was found that the moisture does not influence the transport properties of air at

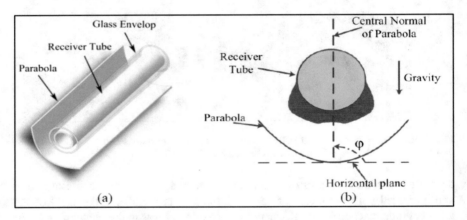

Fig. 3.104 Computational model: (**a**) Schematic of PTC model and (**b**) NUHF profile on the tube wall Li et al. (2015). (License Number: 4615760389745)

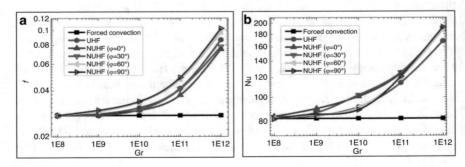

Fig. 3.105 (**a**) Friction factor and (**b**) Nusselt number versus Gr number at Re = 2×10^4 Li et al. (2015). (License Number: 4615760389745)

Fig. 3.106 Cross section
of the heating element
Basbous et al. (2015)

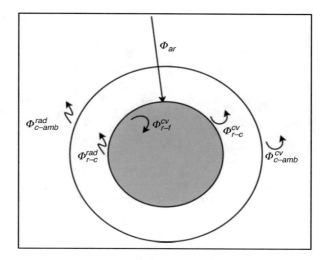

low temperature levels. But, they changed, at high temperatures, with the water vapor amount except the thermal expansion. These changes increased the convective heat losses in the annulus at low humidity and become closer to the dry air at high humidity. It was observed the radiation heat losses in the annular space decreased by the moisture. This study has recommended that extra experiments are needed to obtain reliable results.

Li et al. (2016) studied numerically the impact of the free convection of superheated steam in a PTC receiver tube on the thermal and flow fields as shown in Fig. 3.107. The thermal and flow fields were analysed for uniform (UHF) and nonuniform heat fluxes (NUHF) with wide range of Grashof, Reynolds numbers and solar elevation angles. The outcomes revealed that the free convective boosted the thermal field by more than 10% when the Grashof number is greater than a threshold value. It was concluded that the buoyancy force created a high-speed region at the bottom section of the absorber pipe. The vortex strength under NUHF was greater than that under UHF. They suggested empirical correlations for fRe and Nu number under NUHF to be utilised in practical operation.

Bortolato et al. (2016) used an innovative flat aluminium absorber in small linear solar collector for process heating and direct steam generation system as shown in Fig. 3.108. The absorber width was optimised using a MCRT analysis, and it was constructed using the bar-and-plate technology, including its internal turbulator. It was experimentally placed on an asymmetrical PTC having a concentration ratio of 42. A procedure was applied to examine the PTC during both liquid heating and direct steam generation. The results show that the single- and two-phase flow datasets coincided at decreased temperature. The experimental optical efficiency of this prototype was equal to 82%, while the thermal efficiency at 0.160 K.m²/W was around 64%, with negligible pressure drop. This is a promising value and can still be improved by using flat absorber and low thermal emittance coating material. It was found that the time constant of the collector is equal to 213 s.

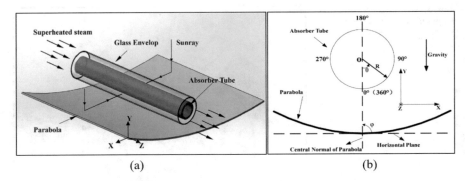

Fig. 3.107 (**a**) Schematic of three-dimensional PTC and (**b**) Cross section of PTC physical model Li et al. (2016). (License Number: 4615781438086)

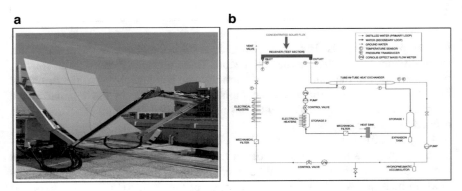

Fig. 3.108 (**a**) prototype of the asymmetrical PTC used in the experimental study and (**b**) Sketch of the experimental rig Bortolato et al. (2016). (License Number: 4615810322976)

Bellos et al. (2017a, b, c, d, e) developed a thermal model to assess the exergetic performance of commercial PTC using Therminol VP1 and air as shown in Fig. 3.109. It was concluded that the thermal oil exhibited high thermal and exergetic effectiveness with 72.49% and 31.57%, respectively. It was noticed that the exergetic efficiency mainly depended on the inlet temperature, while the flow rate is not so paramount. It was pointed out that the exergetic thermal losses were higher in air case due to its lower thermal efficiency. The exergetic frictional losses were higher for air case, which significantly affected the exergy destruction, something was not noticed in thermal oil case.

Bellos et al. (2017a, b, c, d, e) investigated energetically and exergetically a commercial PTC (Eurotrough ET-150) as shown in Fig. 3.110 or a great temperature range from 300 to 1300 K. Pressurised water, Therminol VP-1, nitrate molten salt, sodium liquid, air, carbon dioxide and helium were the examined working fluids. The results proved that the liquid working fluids present higher thermal effectiveness than gas working fluids. The pressurised water was the most suitable working

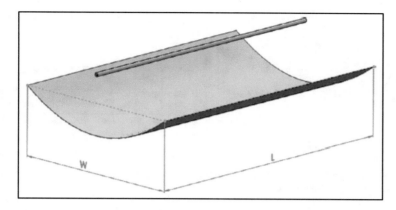

Fig. 3.109 The examined PTC module designed in solidworks Bellos et al. (2017a, b, c, d, e). (License Number: 4616320139237)

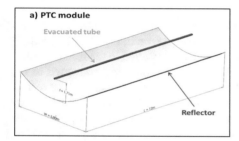 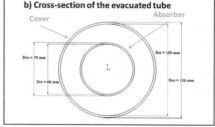

Fig. 3.110 (**a**) PTC module and (**b**) Evacuated tube cross-section Bellos et al. (2017a, b, c, d, e). (License Number: 4616320139236)

fluid for low temperatures up to 550 K, while sodium liquid was the most efficient selection for higher temperatures up to 1100 K. Carbon dioxide and helium perform with similar way and they were the best solutions for extremely high temperatures after 1100 K. The pressure losses have a high impact in gases cases because of its low density, which leads the gases to have optimum exergetic performance in the range of 600–650 K, although they can operate up to 1300 K. The global maximum exergetic performance was observed for sodium liquid case at 800 K and then the exergetic efficiency was equal to 47.48%.

Pavlovic et al. (2017a, b) investigated the impact of PTC's geometric dimensions on the optical, energetic and exergetic efficiencies. The module of the commercial LS-3 PTC was examined with Solidworks Flow Simulation in steady-state conditions as shown in Fig. 3.111. Various combinations of reflector widths and receiver diameters were tested. The optical and the thermal performance, as well as the exergetic performance were calculated for all the examined configurations. The results have shown that higher widths demand higher receiver diameter for optimum performance. For inlet temperature equals to 200 °C with Therminol VP-1, the

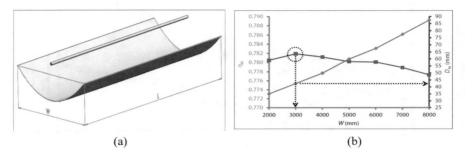

(a)   (b)

Fig. 3.111 (a) The module of LS-2 PTC and (b) Optimum design determination Pavlovic et al. (2017a, b). (License Number: 4615760389821)

optimum design was found to be 3000 mm width with 42.5 mm receiver diameter, using a focal length of 1840 mm. The thermal efficiency and the exergetic efficiency were maximised for a specific receiver diameter, which is fully connected with the width of the collector.

An experimental examination was conducted to assess the energetic and exergetic thermal effectiveness of a PTC receiver tube as shown in Fig. 3.112 by Chafie et al. (2018). The PTC was designed, constructed and installed in the research laboratory at Borj Cedria of Tunisia. The useful energy gain, exergy converted to the fluid, energy and exergy efficiencies and exergy factor were measured. The results have shown that the energy gain converted to the fluid and the energy efficiency were smaller than the useful exergy rate and the exergy efficiency. The DNI and the operating temperature were the main parameters affected the exergy factor. The daily average energy, exergy efficiencies and the exergy factor achieved during time of the experiments were 36.10%, 11.85% and 0.11, respectively.

Kumar and Kumar (2018) investigated experimentally the thermal performance of a PTC having a non-evacuated tube and tested using water under the northern Indian weather conditions as shown in Fig. 3.113. Two cases with and without glazing on the PTC's receiver thermal effectiveness were considered. The results revealed that the PTC's effectiveness relied mainly on the fluid's mass flow rate. The PTC's thermal efficiency varied from 25.84% to 53.55% and 13.8% to 47.5%, respectively, in south-facing mode with and without glazing on the receiver's tube. However, in case of tracking mode, these values were 18.5–33.42% and 12.98–25.72%. Good effectiveness was obtained in south-facing mode compared with tracking mode case and the effectiveness enhanced when using glazing on the receiver's tube. There was no noticeable increase in the PTC's efficiency when the mass flow reached a value of 0.024 kg/s.

Fan et al. (2018) proposed a PTC receiver having twin glass tube and evaluated its effectiveness by a mathematical model as shown in Figs. 3.114 and 3.115. Strategies including replacing the selective coating with non-selective one, lowering the fluid's velocity and lowering the inner glass tube emissivity were investigated. The proposed receiver was suggested to be coated with black paint coating having absorptivity (α) of 0.96, and emissivity (ε) of 0.78. The fluid's velocity had an

Fig. 3.112 (**a**) The experimental system and (**b**) A view of the experimental apparatus Chafie et al. (2018). (License Number: 4616321135118)

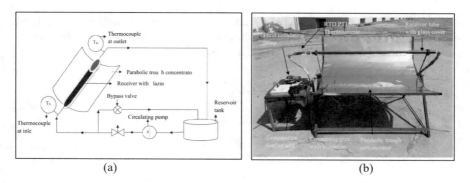

Fig. 3.113 (**a**) The designed hydraulic cycle of experimental system and (**b**) A view of the experimental apparatus Kumar and Kumar (2018). (Order Number: 501495944)

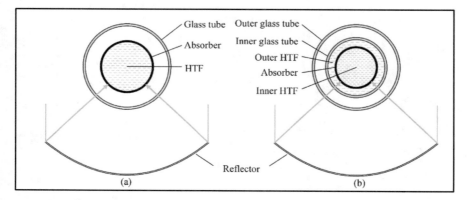

Fig. 3.114 Sketches for a cross section of (**a**) conventional receiver and (**b**) novel receiver Fan et al. (2018). (License Number: 4616330150735)

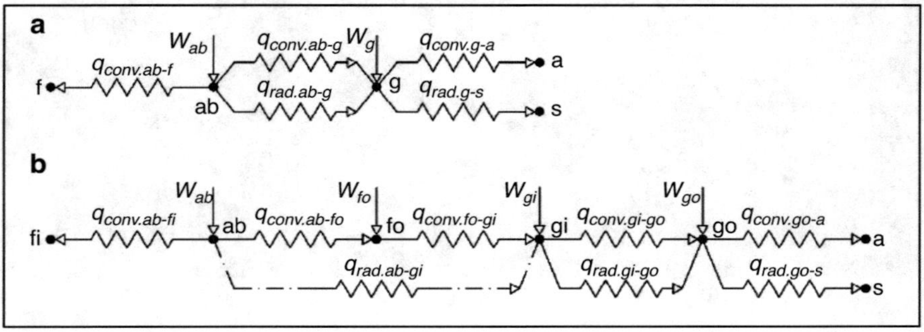

Fig. 3.115 Thermal model of (**a**) conventional receiver and (**b**) proposed receiver Fan et al. (2018). (License Number: 4616330150735)

extensive impact on the PTC's effectiveness than the transmittance of fluid. Furthermore, the usage of selective coating on the inner glass tube was found beneficial at inlet temperature not higher than 150 °C.

Chapter 4
PTC Enhancement Using Nanofluids

4.1 Introduction

One potential answer for enhancing the thermal productivity of PTCs is the utilisation of nanofluids as working fluids. Actually, it is sensible to anticipate an expansion in the thermal productivity of PTCs when the heat transfer base liquid is subbed with proper grouping of nanoparticles. Utilising nanofluid as a working liquid in sun-based bodies is a novel way to deal with increment the galaxies proficiency. One of the valuable uses of nanofluids is in the absorption sunlight-based collectors where the sun pillars are straightforwardly consumed by the nanofluid. Considering the way that most base working liquids utilised in immediate absorption sunlight-based thermal collectors, e.g., water, oil and ethylene glycol, have low absorption coefficients, it follows that expansion of nanoparticles to them upgrades their optical properties as well as enhance the PTC's productivity too (Menbari et al., 2017).

4.2 Nanofluids

Expanding the thermal conductivity of the working fluid would improve the heat transfer rate. Metallic particles, metallic oxides and nanotubes have higher thermal conductivity than that of fluids. For instance, the thermal conductivity of copper at room temperature is around multiple times more than that of water and around 3000 times more than that of oil. Expansion of fine particles (1–100 nm) into heat transfer base-liquids (i.e., nanofluids) can fundamentally boost the heat transfer rate. Clearly, the exploration on the utilisations of nanofluids have been promoted during the ongoing years. The purported nanofluids, a term that was first presented by Choi in 1995 at the Argonne National Laboratory, U.S.A (Choi, 1995). It is likewise characterised, as a blend comprises from an ordinary liquid, e.g., water, oil, ethylene

© The Author(s), under exclusive license to Springer Nature Switzerland AG 2023
H. A. Mohammed et al., *Parabolic Trough Solar Collectors*,
https://doi.org/10.1007/978-3-031-08701-1_4

glycol, glycerol, with a limited quantity of metallic or metallic oxide nanoparticles. Nanofluids have enhanced the liquid thermophysical properties, e.g., thermal conductivity, viscosity and heat transfer coefficients, contrasted to ordinary liquids.

The most common nanoparticles are: Al_2O_3, CuO, TiO_2, SiO_2, Fe_2O_3, ZnO and Au. The nanofluid was deemed as the new age of cutting-edge heat transfer liquids or a two-stage framework which utilised for different designing and modern applications because of its incredible execution. A portion of these applications including atomic reactors, transportation industry, mechanical energy, cooling of microchips, improving diesel generator effectiveness, sunlight-based absorption, microelectronics, biomedical fields and numerous other applications. Nanofluids are additionally called as extraordinary coolant liquids because of their high capacity to assimilate heat more than any customary liquids, so they can diminish the size of frameworks and boost its proficiency. The utilisation of these nanoparticles prompts higher thermal conductivity and subsequently higher heat transfer rate to the working liquid in PTC system. Higher heat transfer rate, which implies higher heat transfer coefficient prompts a lower temperature in the PTC and to bring down heat losses, the way that prompts higher thermal effectiveness (Bellos et al., 2017a). Thus, it is basic to decide the heat transfer and the effectiveness upgrade in PTCs with precision to assess the usage of nanofluids in such frameworks. Therefore, the utilisation of nanofluids is considered among the most effective working liquids as indicated by the literatures, which have been summarised below in the following sections.

Khullar et al. (2012) used sun-based energy via the utilisation of nanofluid in a PTC as shown in Fig. 4.1. Finite difference technique was utilised to solve the governing equations to theoretically analyse the PTC performance. The outcomes were validated against the available experimental data of classical PTC under similar conditions. The results revealed that the PTC's thermal efficiency has 5–10% higher compared to the classical PTC using similar external conditions. It was obviously indicated that the PTC based nanofluids has the ability to exploit sun-based energy more effectively than a classical PTC.

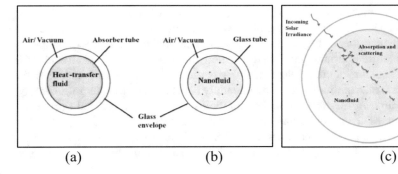

Fig. 4.1 Schematic of (**a**) conventional receiver, (**b**) PTC and (**c**) solar irridiance passing through the nanofluid shows combined heat losses Khullar et al. (2012). (License Number: 4615760388823)

Kasaeian et al. (2012) studied numerically turbulent flow mixed convective of Al_2O_3-oil nanofluid in a PTC pipe as shown in Fig. 4.2. The used various ranges of nanoparticle volume fraction with less than 5%, and operational temperature of 500 K and pressure of 20 psig. The outcomes revealed that the Nu number has direct reliance on the nanofluid's volume fraction. The heat transfer decreased by increasing the operational temperature at a given mass flowrate.

de Risi et al. (2013) suggested a novel solar Transparent PTC using gas-based nanofluids (mixture of CuO and Ni nanoparticles) as shown in Fig. 4.3. They developed a thorough mathematical model considering various aspects of the TPTC and used it to run an optimisation procedure of TPTC. In addition, a genetic algorithm optimisation (MOGA II) was utilised to improve and optimise the solar collector's performance. The results revealed that TPTC based gas-nanofluids can be a promising way compared to classical systems which utilise synthetic oils or molten salts. The optimisation procedure found a maximum thermal efficiency of 62.5% at

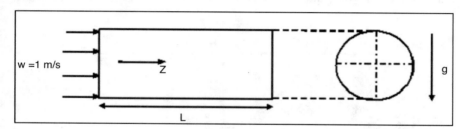

Fig. 4.2 Absorber tube diagram Kasaeian et al. (2012). (License Number: 4615760389567)

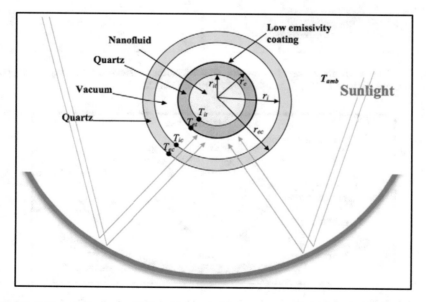

Fig. 4.3 Schematic for a solar PTC de Risi et al. (2013). (License Number: 4616351112046)

nanoparticles volume fraction of 0.3% and the minimum solar radiation to run the plant is 133.5 W/m² and under this value the power plant has to be stopped.

Sokhansefat et al. (2014) studied numerically 3D mixed convective of Al₂O₃-oil nanofluid in a PTC tube subjected to a NUHF as shown in Fig. 4.4a. The effects of various Al₂O₃ particle concentrations, and operational temperatures of 300–500 K on the thermal field were investigated. The heat flux in the circumferential direction was obtained using the MCRT technique. The numerical results show that the heat transfer has a direct dependency on the nanoparticles volume fraction as shown in Fig. 4.4b. In addition, the heat transfer enhancement reduced as the absorber operational temperature increased. The optimum heat transfer was achieved at θ = 310° and θ = 230° (tube's left and right sides), respectively.

Ghasemi and Ahangar (2014) tested Cu-water nanofluid numerically in a PTC to examine its thermal performance as shown in Fig. 4.5. The thermal field and the thermal efficiency of nanofluids based PTC were evaluated and compared with the classical PTC. Further, the influences of different factors as the mass flow rate,

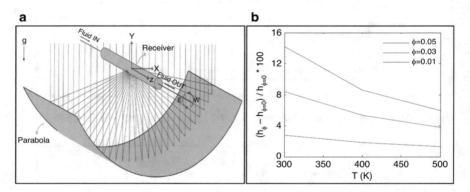

Fig. 4.4 (a) PTC and its absorber tube and (b) convection heat transfer coefficient versus temperature Sokhansefat et al. (2014). (License Number: 4616351253239)

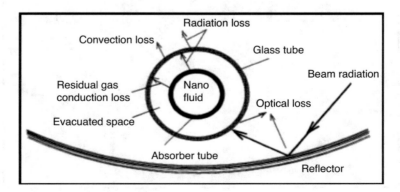

Fig. 4.5 Schematic diagarm of nanofluid-based PTC Ghasemi and Ahangar (2014)

nanoparticle concentration, receiver's geometrical parameters were also investigated. The outcomes show that the addition of limited quantity of copper nanoparticles into water enhanced significantly its absorption characteristics. The thermal and optical efficiencies were improved, and outlet temperatures were higher. The nanofluid based linear PTC has a higher thermal efficiency than a similar conventional collector.

Sunil et al. (2014) investigated experimentally the PTC performance using SiO_2-H_2O based nanofluid as appeared in Fig. 4.6a, b with nanoparticle volume fraction in the range of 0.01–0.05%. The results revealed that SiO_2-H_2O nanofluid has relatively higher thermal efficiency at high velocities and the instantaneous efficiency increased with the increase in velocity. The maximum instantaneous efficiency obtained for SiO_2-water nanofluid for volume flow rate of 20, 40 and 60 L/h is 10.45%, 21.55% and 30.48%, respectively. The corresponding maximum thermal efficiency is 11.27%, 11.61% and 10.95%, respectively. The corresponding maximum overall thermal efficiency is 7.76%, 7.73% and 7.48%, respectively.

Mwesigye et al. (2015) used synthetic oil-Al_2O_3 nanofluid in a PTC to perform a thermodynamic analysis using the entropy generation minimisation method. The PTC used has a rim angle of 80° and a concentration ratio of 86 as shown in Fig. 4.7. The results revealed that the thermal efficiency of the PTC enhanced by up to 7.6% when utilising nanofluids. The entropy generated was a minimum at the optimum

Fig. 4.6 (a) Experimental setup of PTC system and (b) SEM image of SiO_2 nanoparticles Sunil et al. (2014). (License Number: 4615760345973)

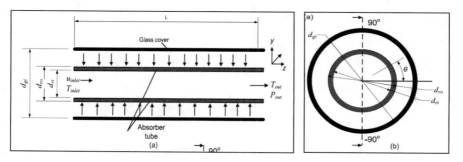

Fig. 4.7 PTC receiver computational domain, (a) longitudinal view and (b) cross-section view Mwesigye et al. (2015). (License Number: 4616380969153)

Reynolds number, at each inlet temperature and nanoparticle concentration, beyond which the utilisation of nanofluids is thermodynamically unreformable. Considerable enhancement in thermal efficiency for all cases was achieved for flow rates lower than 25 m³/h.

Mwesigye et al. (2015) used Syltherm800-CuO nanofluid to examine the thermal effectiveness of a PTC having high concentration ratio of 113 compared the available commercial systems of 82. The results revealed that the receiver's thermal effectiveness decreased with both heat loss and CTD increased when the concentration ratios increased as appeared in Fig. 4.8. The outcomes also show that the utilisation of nanofluids remarkably enhanced the receiver's thermal effectiveness. The thermal effectiveness and thermal efficiency increased up to 38% and 15%, respectively. Practical correlations for Nu number and friction coefficient were obtained and presented.

Zadeh et al. (2015) focused on the development of an efficient modelling and optimisation of a PTC absorber tube subjected to a NUHF using Al₂O₃/synthetic oil as shown in Fig. 4.9. A genetic algorithm (GA) optimisation method and sequential quadratic programming (SQP) were used in the analysis. The Nu number, pressure drop with Re and Ri numbers were used in the optimisation problem as design constraints. The results have shown that the thermal performance of the absorber tube is improved when nanoparticles volume fraction increased. The thermal effectiveness decreased as the absorber's pipe inlet temperature increased. The hybrid optimisation algorithm (GA-SQP) has provided an efficient technique of obtaining good solution and reducing the calculation time using SQP.

Chaudhari et al. (2015) investigated experimentally using Al₂O₃-water on the thermal effectiveness of a PTC as shown in Fig. 4.10. The nanoparticles used have 0.1% of volume fraction and 40 nm particles dimension. The outcomes illustrated that nanofluid-based PTC has higher thermal effectiveness than water-based PTC. It was revealed that the thermal efficiency increased by 7% at 0.1% of nanoparticle volume fraction with heat transfer augmentation of 32% when using nanofluid.

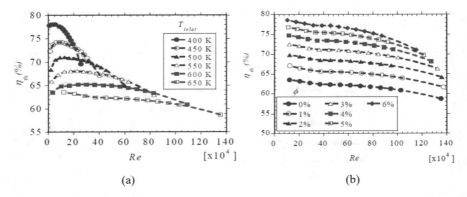

Fig. 4.8 (a) Influence of inlet temperature and (b) volume fraction on the PTC's thermal efficiency Mwesigye et al. (2015). (License Number: 4616380969157)

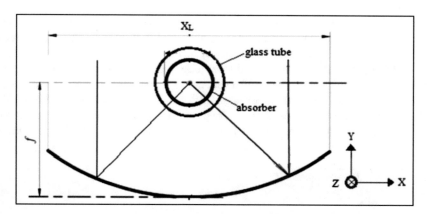

Fig. 4.9 Schematic diagram of PTC and its absorber tube Zadeh et al. (2015). (License Number: 4616880495624)

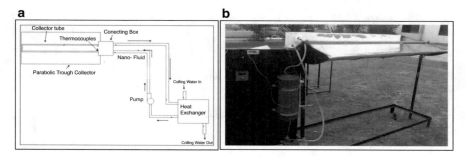

Fig. 4.10 (a) The experimental setup diagram and (b) The experimental apparutus Chaudhari et al. (2015). (License Number: 4616880495626)

Basbous et al. (2015) studied numerically the possibility of PTC installation in Casablanca of Morocco. They studied the influence of moisture on the PTC design and the heat transfer with the existence of water vapor. The influence of humidity on the humid air thermo-physical properties was assessed to evaluate the heat transfer. The optical properties of water vapor were determined for the sake of taking the contribution of humid air in radiative heat transfer calculations. The results revealed that the moisture does not have any impact on the transport properties of air at low temperature levels. However, they varied significantly, at high temperature levels, with the amount of water vapor except the thermal expansion. These changes increased the convective heat losses in the annulus when the humidity is low, and they become closer to those of dry air when the humidity is high. It was also found that the moisture decreased the radiation heat losses in the annular space and the thermal effectiveness was impacted. It was stated that further experiments are needed to obtain reliable results.

Kasaeian et al. (2015) constructed a standard pilot of trough collector having a 0.7 m width and 2 m in height reflector made of steel mirror to investigate the ways of its performance enhancement as shown in Figs. 4.11 and 4.12. Four kinds of receivers: a black painted vacuumed steel tube, a copper bare tube, a glass enveloped non-evacuated copper tube and a vacuumed copper tube were used to compare their optical and thermal effectiveness. Multi walled carbon nanotube (MWCNT)/ oil nanofluid with 0.2–0.3% volume fraction were utilised in vacuumed copper absorber tube as shown in Fig. 4.13. The outcomes revealed that the vacuumed

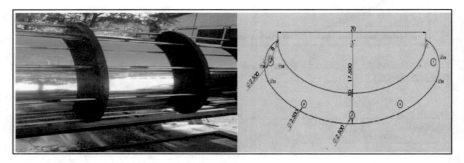

Fig. 4.11 Dimensions of PTC absorber tube Kasaeian et al. (2015). (License Number: 4616921378646)

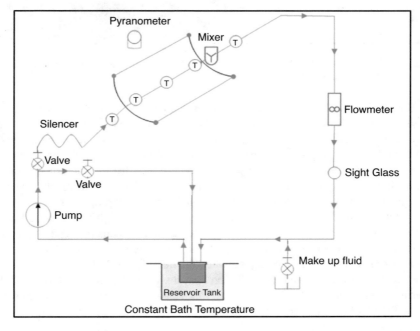

Fig. 4.12 Schematic diagram of the hydraulic cycle Kasaeian et al. (2015). (License Number: 4616921378646)

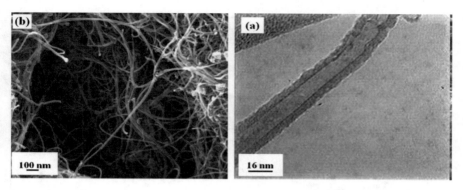

Fig. 4.13 SEM (left) and TEM (right) of the multiwall carbon nanotubes Kasaeian et al. (2015). (License Number: 4616921378646)

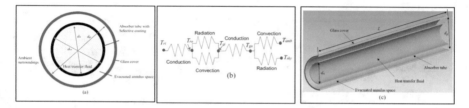

Fig. 4.14 PTC receiver physical model, (**a**) cross-section, (**b**) thermal circuit and (**c**) computational area Mwesigye et al. (2016a, b). (License Number: 4616930900295)

tube's global efficiency is 11% more than the bare tube. The maximum optical and thermal effectiveness of the vacuumed copper tube were 0.61 and 0.68, respectively. The global efficiency is improved by 4–5% and 5–7%, at 0.2% and 0.3% nanoparticle concentration, respectively. The nanofluid shows high thermal potential for further examinations.

Basbous et al. (2016) utilised ultrafine particles Al_2O_3, Cu, CuO and Ag dispersed in Syltherm 800 to examine numerically the thermal effectiveness of a PTC framework. The results revealed that the PTC's heat loss factor declined when using nanofluids. The silver nanoparticles exhibited the optimum PTC's thermal effectiveness with 36% increase in heat transfer coefficient and 21% decrease in heat loss factor.

Mwesigye et al. (2016a, b) presented numerically the thermodynamic effectiveness of a PTC using Cu-Therminol VP-1 nanofluid as shown in Fig. 4.14. The outcomes revealed that the thermal efficiency of the system increased by 12.5% and the entropy generation rates reduced significantly as the nanoparticle volume fraction increased. It was observed that the entropy generation rates reduced between 20% and 30% as the nanoparticle concentration increased from 0% to 6% at flow rates lower than 45 m³/h.

Wang et al. (2016) implemented a combined optical-thermal-stress simulation model based on FEM to examine the PTC effectiveness as shown in Fig. 4.15. The Al_2O_3/synthetic oil nanofluid with NUHF profile were utilised. The effects of

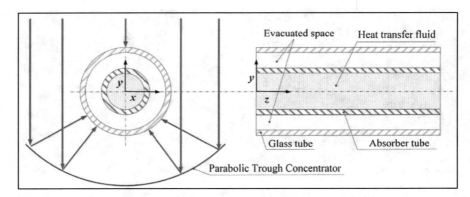

Fig. 4.15 Schematic diagram of the PTC Wang et al. (2016). (License Number: 4616931088244)

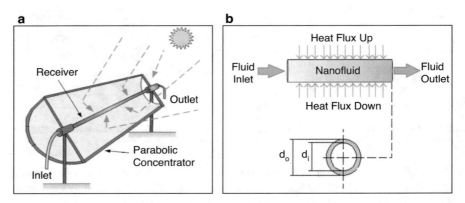

Fig. 4.16 (**a**) Solar PTC system and (**b**) PTC receiver Ghasemi and Ranjbar (2016). (License Number: 4616940175748)

particle concentration and other key factors of the PTC system were also investigated. It was found that the utilisation of nanofluid dramatically diminished the temperature gradients, the maximum temperature and deformation in the absorber especially in y-direction. The y-direction displacements decreased from 2.11 mm to 0.54 mm when the volume fraction increased, this indicates that the new structure solar receiver has a better performance. The new PTC structure has higher thermal efficiency using Al_2O_3/synthetic oil nanofluid contracted to the one using synthetic oil. Moreover, the changes of temperature in the absorber with the direct normal irradiance (DNI), the inlet temperature and the inlet velocity were remarkably reduced.

Ghasemi and Ranjbar (2016) simulated numerically forced convective turbulent nanofluid flow in a solar PTC receiver as shown in Fig. 4.16. A commercial CFD code was employed to find hydrodynamic and heat transfer coefficients by means of Finite Volume Method (FVM). The effect of nanoparticles volume fraction (ϕ) on the PTC thermal effectiveness was studied. The results revealed that by increasing the nanoparticle volume fraction, the average Nusselt number increased for both CuO and Al_2O_3 nanofluids. Furthermore, it was found that the heat transfer coefficient enhanced up to 28% and 35% when utilising Al_2O_3-water and CuO-water nanofluids ($\phi = 3\%$), respectively. On the other hand, the friction factor was found lower using nanofluid compared with water.

Abid et al. (2016) conducted energy and exergy analyses using nanofluids and molten salts for of two kinds of solar collectors used with the steam power plant as shown in Figs. 4.17 and 4.18. These two types were Parabolic dish (PD) and parabolic trough (PT) to obtain solar energy using four fluids, i.e., Al_2O_3, Fe_2O_3 nanofluids, LiCl-RbCl and $NaNO_3$-KNO_3 molten salts. The operating parameters such as solar irradiation and ambient temperature varied to comprehend their effects on the receiver's outlet temperature, overall energy and exergy efficiencies of PD and PT. The outcomes revealed that the outlet temperature of PD was higher than PT under similar working situations due to the high concentration ratio of PD. The overall exergy efficiency of PD and PT was 20.33–23.25% and 19.29–23.09%, respectively, when the ambient temperature increased from 275 K to 320 K. It was noticed that the net power produced by PD and PT and the energetic and exergetic efficiencies were higher using nanofluids than that when using molten salts. It was observed that the PD overall performance was higher when utilising nanofluids compared with PT.

Kaloudis et al. (2016) used CFD numerical model to study the PTC's thermal effectiveness utilising Syltherm 800/Al_2O_3 nanofluid with concentrations (0–4%) as shown in Fig. 4.19a, b. To address the nanofluid-modelling problem the two-phase

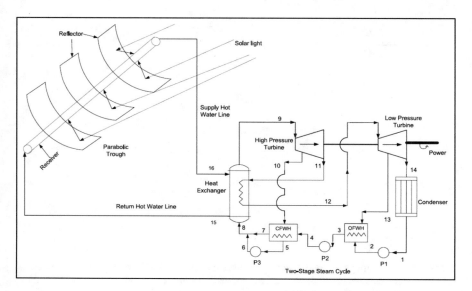

Fig. 4.17 Schematic of PTC integrated with two-stage steam cycle Abid et al. (2016). (License Number: 4616940175755)

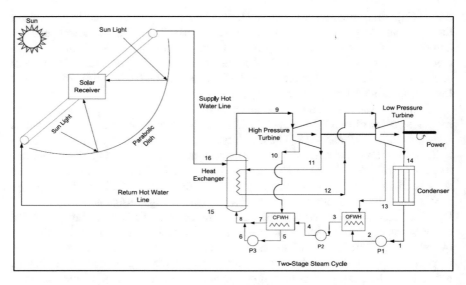

Fig. 4.18 Schematic of PTC integrated with two-stage steam cycle Abid et al. (2016). (License Number: 4616940175755)

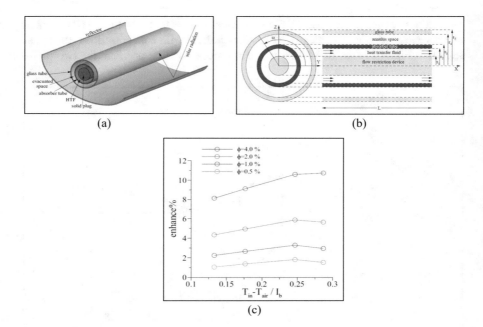

Fig. 4.19 (a) PTC physical model and its absorber tube, (b) LS2-PTC computational model and (c) Percentage increase in efficiency against inlet temperature for various volume concentrations Kaloudis et al. (2016). (License Number: 4616941148672)

approach was preferred (against single-phase model) and validated against experimental and numerical results. In addition, the temperature and velocity fields of the Syltherm 800/Al_2O_3 nanofluid were associated with the enhanced heat transfer occurring at higher nanoparticle concentrations. A boost up to 10% on the collector efficiency was reported for Al_2O_3 concentration of 4% as shown in Fig. 4.19c. It was concluded that the two-phase approach was more accurate compared with the single-phase approach which is more common in similar studies. This increase was associated with detailed temperature and velocity fields, showing enhanced mixed convection effects for higher nanoparticle concentrations.

Coccia et al. (2016) analysed the yearly yield assessment of a low-enthalpy PTC using numerical modelling. Six water-based nanofluids at various volume fractions were investigated: Fe_2O_3, SiO_2, TiO_2, ZnO, Al_2O_3 and Au. The results show that small improvements are associated with Au, TiO_2, ZnO and Al_2O_3 nanofluids at low volume fractions compared with water. However, there was no advantage with respect to water when increasing the nanoparticles concentration as appeared in Fig. 4.20. The improvements in thermal efficiency were related to low concentrations of nanoparticles. This is because of the dynamic viscosity tends to considerably increase with the nanoparticle concentration. However, since the dynamic viscosity decreased with temperature while the thermal conductivity increased, it could be of some interest evaluating the potential of nanofluids at higher temperatures. This could be achieved by conducting further experimental investigations at high temperatures on reduced nanoparticle concentration such as TiO_2, ZnO, Al_2O_3 and Au, which did not give negative results at the temperature range investigated in this study.

Ferraroa et al. (2016) analysed the behaviour of a PTC operating with nanofluids, and compared its performance to the more traditional ones using oil as shown in Fig. 4.21. A thermal analysis model was developed and implemented using Matlab. The simulations were performed for a suspension of Al_2O_3 in synthetic oil and its characteristics compared to the corresponding base-liquid. The simulations were

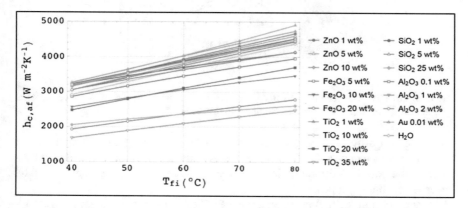

Fig. 4.20 Convective heat transfer coefficient $h_{c,af}$ as a function of the inlet fluid temperature T_{fi} Coccia et al. (2016). (License Number: 4616941484393)

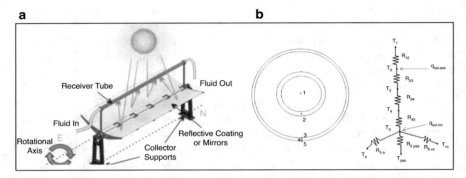

Fig. 4.21 (a) PTC system and (b) Analogue electrical cicuit of a PTC Ferraroa et al. (2016). (License Number: 4616941484382)

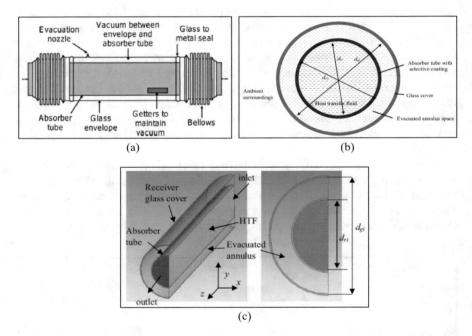

Fig. 4.22 (a) PTC receiver tube, (b) Receiver's cross-sectional view and (c) Receiver's lateral view Mwesigye et al. (2018). (License Number: 4616950131106)

carried out for different DNI and variable mass flow, ensuring a temperature at the collector outlet below 400 °C. It was observed that there were only slight differences in the power loss and efficiency, while the main advantage is represented by lowering the pumping power.

Mwesigye et al. (2018) investigated the PTC's optimum thermal and thermodynamic effectiveness utilising Cu-Therminol, Ag-Therminol and Al_2O_3-Therminol nanofluids as shown in Fig. 4.22. They used finite volume method with the aid of CFD tool together with MCRT method. The outcomes revealed that the highest and

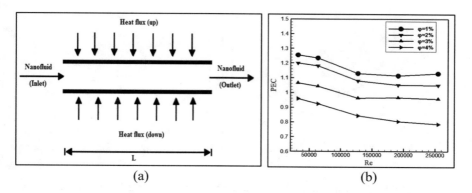

Fig. 4.23 (a) Schematic of nanofluid flowing in a PTC receiver tube and (b) PEC versus Re number Ghasemi and Ranjbar (2017a, b). (License Number: 4616950280403)

lowest thermal effectiveness was obtained using Ag-Therminol and Al$_2$O$_3$-Therminol, respectively. It was found that the thermal efficiency increased by 13.9%, 12.5% and 7.2% for Ag-Therminol, Cu-Therminol and Al$_2$O$_3$-Therminol, respectively, at a proportion ratio of 113. The maximum thermodynamic effectiveness for low exergy destruction was essentially reliable on the inlet temperature used. Empirical correlations were developed to provide improved thermodynamic performance.

Ghasemi and Ranjbar (2017a, b) simulated numerically 3D turbulent nanofluid flow and heat transfer inside a PTC receiver tube as shown in Fig. 4.23a. CFD simulations with the aid of Finite Volume Method (FVM) were carried out to examine the impact of using nanofluid on the PTC's thermal efficiency. The results indicated that, using of nanofluid instead of base fluid as a working fluid causes an increase in the effective thermal conductivity and leads to thermal enhancement. Furthermore, the results revealed that by increasing of the nanoparticle volume fraction, the performance evaluation criteria is increased as shown in Fig. 4.23b.

Bellos et al. (2017a, b, c, d, e) investigated the use of nanofluids in PTC for various cases as shown in Fig. 4.24a. They used nanoparticles (Al$_2$O$_3$ and CuO) suspended in Syltherm 800 oil. A detailed thermal model was developed in EES, and its results were validated with the experimental results. The results found that the utilisation of nanofluids increased the PTC's effectiveness about 0.5%. The Nusselt number enhancement was about 50% and it was the remarkable for CuO nanoparticles as shown in Fig. 4.24b. It was affirmed that high temperature values produced higher thermal effectiveness and it can reach up to 0.6%. Additionally, it was observed that the nanofluids with high concentration and low flow rates provided higher enhancement, which would be paramount criteria to select nanofluids in PTC frameworks.

Khakrah et al. (2017) performed a comprehensive numerical analysis of 3D turbulent flow of Al$_2$O$_3$/synthetic oil nanofluid in a PTC receiver tube as shown in Fig. 4.25. Various governing parameters such as the wind velocity, nanoparticles concentration, inlet temperature and reflector's angle were investigated. The results

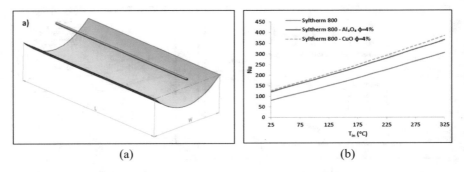

(a)   (b)

Fig. 4.24 (**a**) PTC module and (**b**) Nusselt number comparison for the initial case study Bellos et al. (2017a, b, c, d, e). (License Number: 4619140019400)

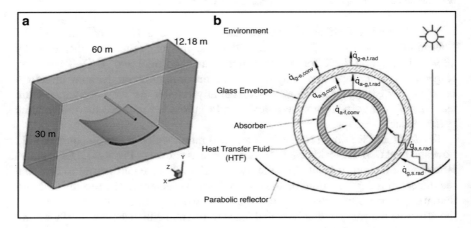

Fig. 4.25 (**a**) Computational domain and (**b**) Reflector and receiver parts Khakrah et al. (2017). (License Number: 4616931214653)

revealed that the reflector's orientation significantly affected the Nu number values around the receiver tube. It was observed that the convection heat loss increased by 123.2%, 106.5%, 95.7% and 86.1% for wind velocities of 5, 10, 15 and 20 m/s, respectively, when using reflector surface with 30° rotation compared with the flow over horizontal reflector. The PTC's efficiency diminished by 8% when the temperature difference (inlet and ambient) was doubled. The efficiency enhancement was 14.3% when 5% volume fraction of Al_2O_3 added to synthetic oil. The wind velocity has no influence on the efficiency enhancement for the rotated reflector cases.

Kasaeian et al. (2017) presented new forms of PTC as direct absorption solar collectors with three different receivers made of glass: a bare tube, non-evacuated tube and a vacuumed tube as shown in Fig. 4.26. The influences of using nanofluids and the properties of the absorber tube on the PTC's thermal effectiveness were investigated. Two nanoparticles of 0.2% and 0.3% MWCNT and nanosilica dispersed in ethylene glycol were utilised. It was found that the maximum outlet

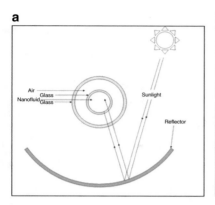
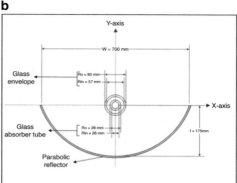

Fig. 4.26 (a) The cross-section of the collector and (b) The dimensions of the reflector and the receiver in X-Y coordination Kasaeian et al. (2017). (License Number: 4620000055072)

temperature and thermal effectiveness in all three receivers was achieved with 0.3% MWCNT/EG. The vacuumed glass-glass receiver exhibited outlet temperature and thermal efficiency of 338.3 K and 74.9%, respectively, which was 20% higher than the bare tube. The carbon nanotubes exhibited maximum values of volume fraction and thermal efficiency at 0.5, and 80.7%, respectively, and it was 0.4 and 70.9%, respectively, for nanosilica.

Mwesigye et al. (2018) investigated numerically using energy and exergy analyses the PTC's thermal effectiveness using SWCNTs-Therminol® VP-1 nanofluid as shown in Fig. 4.27. The results revealed that the thermal effectiveness was augmented about 234% using SWCNT compared with base-fluid. The thermal efficiency raised up to 4.4% as the nanoparticle volume fraction changed from 0 to 2.5%. The thermal efficiency was reasonably improved at flow rates lower than 28 m³/h. It was observed that SWCNT nanofluid increased the exergetic performance of the PTC system. It was pointed out that the entropy generation rate can be reduced up to 70% due to the significant boost in the thermal effectiveness and the decline in the temperature difference. It was also affirmed that the higher thermal efficiency or energy output does not necessarily come from higher thermal conductivity. The specific heat capacity must be considered as another paramount factor in evaluating the PTC's thermal effectiveness utilising nanofluids.

Subramani et al. (2018) investigated experimentally the performance of TiO₂/water nanofluids in a solar PTC under turbulent flow regime as shown in Fig. 4.28. The thermal and flow fields of nanofluids through the collector were investigated. The outcomes revealed that the Nusselt number, absorbed energy parameter and the maximum efficiency enhancement were augmented up to 22.76%, 9.5% and 57%, respectively, when using TiO₂ nanofluids instead of base-fluid. Practical equations were derived for the Nusselt number, friction factor and performance index criterion.

Bellos et al. (2018a, b, c) used different types of nanoparticles such as Cu, CuO, Fe₂O₃, TiO₂, Al₂O₃ and SiO₂ dispersed in Syltherm 800 and tested numerically in a PTC system using EES as shown in Fig. 4.29a. This study covered a flow rate in the range of 50–300 L/min, inlet temperature in the range of 300–650 K and

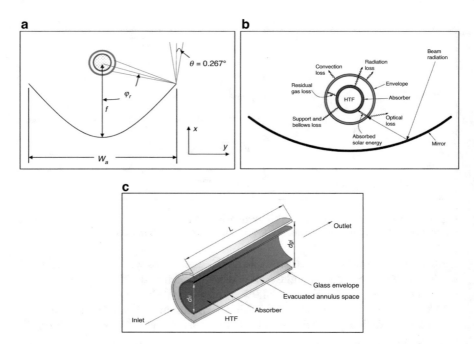

Fig. 4.27 (**a**) The PTC geometry, (**b**) A typical PTC receiver and its corresponding heat losses and (**c**) Computational domain Mwesigye et al. (2018). (License Number: 4620000187041)

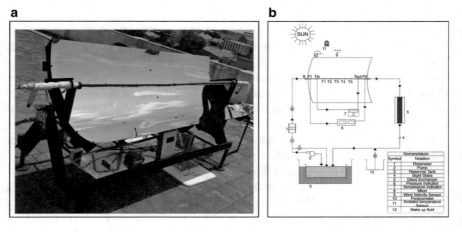

Fig. 4.28 (**a**) Photograph of the PTC and (**b**) Graphical layout of the experimental setup Subramani et al. (2018). (License Number: 4620000338987)

nanoparticle volume fraction up to 6%. The results revealed that the most efficient nanoparticle was the Cu, followed by other nanofluids. It was observed that lower flow rates, higher inlet temperatures and higher nanoparticle concentrations

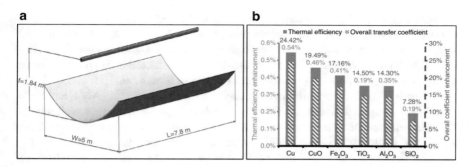

Fig. 4.29 (a) The PTC examined module and (b) Thermal efficiency enhancement and overall heat transfer coefficient Bellos et al. (2018a, b, c). (License Number: 4620010604035)

provided higher improvement, whilst it is relatively unchanged for various solar irradiation levels. The thermal efficiency improvement was 0.31, 0.54 and 0.74% for Cu volume fractions 2, 4 and 6%, respectively, as shown in Fig. 4.29b They suggested a new index, which considered the heat transfer coefficient, the density-specific heat capacity product and the flow rate for the assessment of the thermal performance enhancement.

Bellos et al. (2018a, b, c) examined the addition of CuO nanoparticles in Syltherm 800 and in nitrate molten salt and tested its efficacy on a PTC thermal performance. They also evaluated the PTC total effectiveness by considering the hydraulic analysis (pressure losses) and the exergetic analysis. The outcomes revealed that the mean thermal enhancement with the utilisation of Syltherm 800-CuO was 0.65% compared to pure Syltherm, whilst it was 0.13% with the utilisation of molten salt-CuO. The hydraulic analysis affirmed that the pressure drop increased around 50% for Syltherm 800-CuO compared to the pure Syltherm 800 operation, while it was increased around 16% compared to the pure molten salt for molten salt-CuO case. The exergetic analysis confirmed that the molten salts produced higher exergetic performance compared to the oils. It was observed that, for the molten salt case, the maximum exergetic efficiency was achieved at inlet temperature of 650 K and the exergetic efficiency was around 38.4%.

Marefati et al. (2018) analysed the optical and thermal analyses of PTC. Four cities of Iran with different weather conditions were chosen as case studies to assess the PTC effectiveness. Effective parameters such as concentration ratio, incident angle correction factor, collector mass flow rate were considered. Numerical modelling of the analysis was done using MATLAB software. The results revealed that Shiraz, with an average annual thermal efficiency of 13.91% and annual useful energy of 2213 kWh/m², is the best region to use solar concentrator systems. The utilisation of various nanofluids (CuO, Al$_2$O$_3$ and SiC) with volume fraction of 1% to 5% on the PTC's effectiveness was investigated. It was found that the thermal effectiveness using CuO nanofluid is higher than that other nanofluids. The results of this study were useful for the design and implementation of solar systems as the PTC performance changes with geographic region.

Rehan et al. (2018) evaluated the experimental effectiveness of a locally developed PTC system having a concentration ratio of 11 for domestic heating applications. Two types of nanofluids were utilised and prepared Al_2O_3/H_2O and Fe_2O_3/H_2O at different concentrations. The experiments were conducted at Taxila of Pakistan under wide range of working parameters as shown in Figs. 4.30, 4.31, and 4.32. The results revealed that the maximum efficiencies were obtained using Al_2O_3 and Fe_2O_3 nanofluids compared to water. It was found that Al_2O_3 nanofluids were preferable in the efficiency enhancement compared to Fe_2O_3 for domestic applications utilising PTC. The results provided essential vision from the commercialisation point of view for developing in-house linear PTC and the impact of utilising nanofluids for space heating applications.

De los Rios et al. (2018) investigated experimentally the impact of Al_2O_3-water nanofluid on the PTC thermal effectiveness as shown in Fig. 4.33. The results revealed that the PTC thermal efficiency was significantly enhanced for all nanofluids examined with relying on the incident angle value. The maximum efficiency of 52.4% was achieved when the incident angle varied from 20° to 30° (as shown in Fig. 4.34), whereas it was 40.8% when using water. The maximum efficiency was 57.7% when using an incident angle of 10°, and nanofluid with 1% concentration, while it was 46.5% when using water. The PTC's outlet temperatures were higher using nanofluids than those when using water even when the solar radiation value was very small.

Allouhia et al. (2018) proposed a 1D- mathematical model to study the influence of various nanoparticles in a PTC system as shown in Fig. 4.35. Energetic and exergetic analyses were done using various nanoparticles Al_2O_3, CuO and TiO_2 with different volume fraction. It was revealed that the existence of nanoparticles enhanced the thermal effectiveness and produced higher values of Figure of Merit

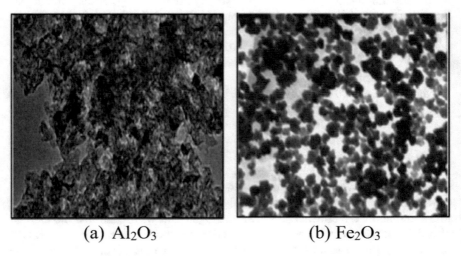

(a) Al_2O_3 (b) Fe_2O_3

Fig. 4.30 TEM micrograph of Al_2O_3 and Fe_2O_3 nanoparticles Rehan et al. (2018). (License Number: 4620010779151)

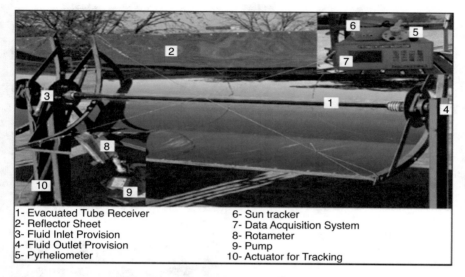

1- Evacuated Tube Receiver	6- Sun tracker
2- Reflector Sheet	7- Data Acquisition System
3- Fluid Inlet Provision	8- Rotameter
4- Fluid Outlet Provision	9- Pump
5- Pyrheliometer	10- Actuator for Tracking

Fig. 4.31 Experimental setup of PTC with evacuated tube receiver Rehan et al. (2018). (License Number: 4620010779151)

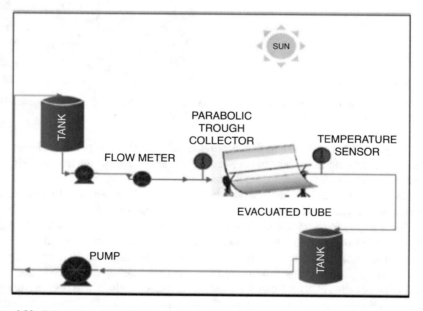

Fig. 4.32 Schematic view of experimental setup Rehan et al. (2018). (License Number: 4620010779151)

(FoM). The FoM value, for CuO nanofluid, was greater than 1 at nanoparticle volume fraction>1% and it exceeded 1.8 when the volume fraction was 5%. It was observed that CuO nanofluid increased remarkably the outlet temperature

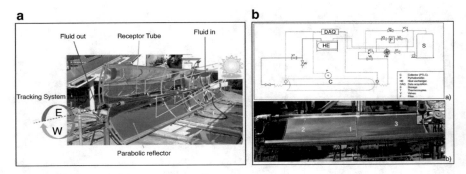

Fig. 4.33 (**a**) Experimental system for linear PTC and (**b**) PTC model De los Rios et al. (2018). (License Number: 4620020752103)

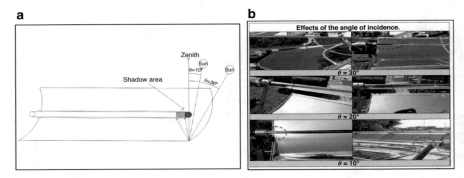

Fig. 4.34 (**a**) Presentation of the incident angle in PTC and (**b**) Effect of incident angles of the PTC De los Rios et al. (2018). (License Number: 4620020752103)

contrasted to other nanofluids, which provided similar thermal pattern. The exergy efficiency, for base fluid, ranged between 3.05% and 8.5%, whereas it was significantly enhanced to 9.05% for CuO nanofluid.

Alsaady et al. (2018) used ferrofluids experimentally to test its efficacy on the thermal effectiveness of a small-scale solar PTC as shown in Fig. 4.36. Two working fluids Fe_3O_4 ferrofluids and distilled water were utilised as shown in Fig. 4.37. The outcomes revealed that the thermal efficiency increased by 16% and 25% with the utilisation of ferrofluids without and with external magnetic field compared to base fluid, respectively. It was observed that the ferrofluids show much better thermal efficiency than classical fluids at high temperature levels. It was noticed that the addition of electromagnetics of 0.01 °C m²/W to the new PTC provided highest thermal efficiency value as it significantly enhanced the ferrofluids thermo-physical properties.

Tagle-Salazar et al. (2018) presented a thermal mathematical model of PTC for heating applications using alumina-water nanofluid with the aid of EES as shown in Fig. 4.38. Furthermore, experimental work was also conducted using a PTC

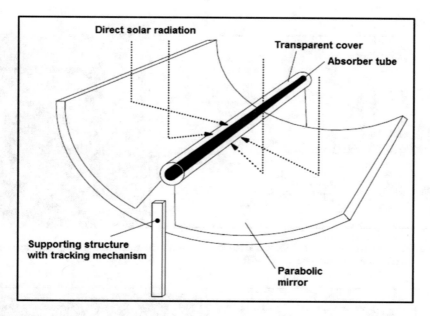

Fig. 4.35 The cross-section of the collector, reflector and receiver tube Allouhia et al. (2018). (License Number: 4620030644919)

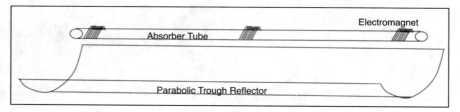

Fig. 4.36 Schematic diagram of PTC system Alsaady et al. (2018). (License Number: 4620030655616)

framework as shown in Fig. 4.39. It was noticed that the variation in the simulation results was less than 1% for outlet temperature and 10–15% for the thermal efficiency. This was attributed to the experimental measurement uncertainties, combined with a small error in the assumptions used in the mathematical thermal model. For one collector, there were small enhancements in both heat gain (+0.3 W/m) and thermal efficiency (+0.03%). It was inferred that the thermal behaviour of the non-evacuated receiver remarkably enhanced with high thermal losses contrasted to an evacuated receiver.

Kumar et al. (2018) studied experimentally solar PTC with a modified absorber tube with copper fins and using TiO_2 nanofluid as shown in Fig. 4.40. The study was investigated various factors such as DNI, inlet velocity and inlet temperature. The

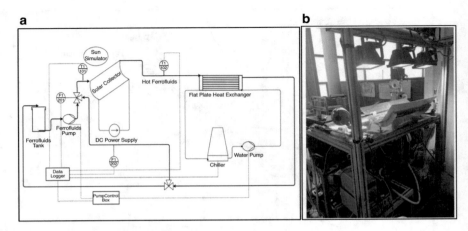

Fig. 4.37 (**a**) The experimental setup and (**b**) Photograph of the experimental setup Alsaady et al. (2018). (License Number: 4620030655616)

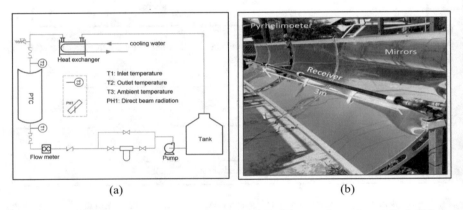

Fig. 4.39 (**a**) The experimental setup and (**b**) Photograph of the experimental setup Tagle-Salazar et al. (2018). (License Number: 4620040411928)

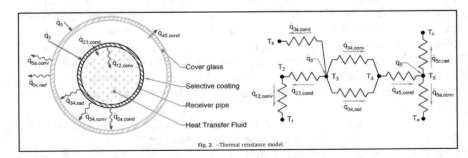

Fig. 4.38 Schematic diagram of the thermal resistance model Tagle-Salazar et al. (2018). (License Number: 4620040411928)

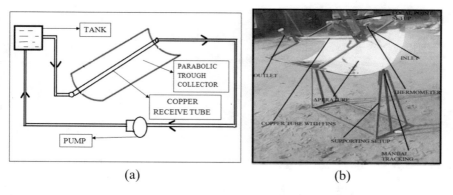

Fig. 4.40 (**a**) Schematic diagram of PTC and (**b**) Photograph of modified PTC with copper fin receiver tube Kumar et al. (2018)

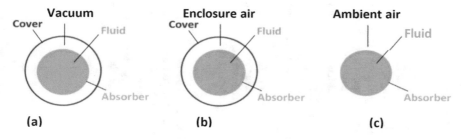

Fig. 4.41 Various configurations of PTC receivers, (**a**) Evacuated tube, (**b**) Non-evacuated tube and (**c**) Bare tube Bellos et al. (2020). (License Number: 4920041356616)

results revealed that the usage of fins increased the thermal effectiveness and the outlet temperature increased with the increases in nanofluid concentration.

Bellos et al. (2020) studied mathematically the thermal augmentation using (Cu/Syltherm 800) nanofluid in a PTC. Three PTC receiver's tube were utilised in the analysis; evacuated, non-evacuated and bare, with no cover as shown in Fig. 4.41. The results revealed that the bare tube has the maximum enhancements and higher thermal losses and nanofluids enhanced its effectiveness. The optimum augmentation for bare tube, nonevacuated tube and evacuated tube was found 7.16%, 4.87% and 4.06%, respectively, when using cermet coating. These augmentations were accordingly found to be 17.11%, 12.30% and 12.24% for the same tube types, respectively, at 25 L/min and when using nonselective receiver.

A summary of literature studies on PTC utilising conventional fluids and mono nanofluids are listed in Table 4.1. It can be obviously noticed from Table 4.1 that most of the research efforts, which have been carried out by different researchers at different countries around the globe, are mainly done using nanofluids with 53% compared with conventional fluids and other fluids with 33% and 14%, respectively, as can be seen in Fig. 4.42. It can also be seen that most studies are performed by Iran using nanofluids with 27%, followed by Greece with 17%, South Africa with

Table 4.1 Summary of literature studies on PTC utilising conventional fluids and mono nanofluids

References	Study type	Heat transfer enhancement technique	Fluid used	Phase mode	Flow type	Remarks	Proposed correlations
Arasu and Sornakumar (2006)	Exprimental	Nil	Water	Single phase	Mass flow rate was in the range of 0.7–1.1 L/min	The PTC has 67 s time constant and has fast response to maintain quasi-steady state conditions. The accuracy of its tracking mechanism was much greater than the required 0.5°	$\eta = 0.6905 - 0.3865\left(\dfrac{\Delta T}{I}\right)$
Cheng et al. (2010)	Numerical	Nil	Syltherm-oil 800	Single phase	Turbulent flow Re number was in the range from 2000 to 6000	The radiation loss in model 3 (the unabridged model that includes all of the three-heat transfer type convection-conduction-radiation) was up to 153.70 W/m². The collector's efficiency enhanced when the radiation loss reduced as low as possible	Nil
Tao and He (2010)	Numerical/ 2D	Nil	Air	Single phase	Turbulent flow Rayleigh number was in the range of $1000–1 \times 10^8$	The effect of natural convection should be considered when Ra number is greater than 10^5. The thermal conductivity had less influence on the nu number and average temperatures when its larger than 200 W/m.K	Nil

Reference	Method		Fluid	Phase	Flow conditions	Findings	
He et al. (2011)	Numerical	Nil	Syltherm 800 oil	Single phase	Not specified	The temperature rise only augmented with CR increase when $\varphi<15°$, and the influence of rim angle on the heat transfer process could be discarded. However, lots of rays were reflected by glass cover, and the temperature rise was much lower when $\varphi>15°$	Nil
Cheng et al. (2012a, b)	Numerical	Nil	Oil syltherm-800, Therminol VP-1, nitrate salt, Hitec XL	Single phase	Turbulent flow Re number was in the range of $3 \times 10^3 - 3.08 \times 10^5$	The thermal oils used have better thermal effectiveness and a lower pressure loss than the two molten salts. The total thermal radiation loss increased when the fluid's inlet temperature increased	Nil
Khullar et al. (2012)	Mathematical 1D/MATLAB	Nil	Therminol VP-1 – Al nanofluid	Single phase	Volume flow rate was in the range of $0.307 \times 10^{-3} - 0.912 \times 10^{-3}$ m³/s	The nanofluid-based PTC had 5–10% higher efficiency than the classical PTC. The solar insolation and incident angle have significant effect on the thermal effectiveness of PTC	Nil
Kasaeian et al. (2012)	Numerical	Nil	Synthetic oil- Al_2O_3 nanofluid	Single phase	Turbulent flow Re number was in the range of 25,000–100,000	The Nusselt numbers had direct reliance on the nanoparticle volume fraction. The thermal characteristics increased as Reynolds number increased	Nil

(continued)

Table 4.1 (continued)

References	Study type	Heat transfer enhancement technique	Fluid used	Phase mode	Flow type	Remarks	Proposed correlations
de Risi et al. (2013)	Mathematical 1 D model with optimisation	Transparent PTC using quartz	Gas-nanoparticle (Ni-CuO) mixture	Single phase	Velocity was in the range of 5–20 m/s	The optimisation procedure provided thermal efficiency of 62.5%. The minimum solar radiation to run the plant was 133.5 W/m². Under this value the power plant has to be stopped	Nil
Hachichaa et al. (2013)	Numerical/CFD	Nil	Thermal oil	Single phase	Mass flow rate was 0.68 kg/s	The developed model was successfully able to determine the heat losses, temperature in the heat collector element (HCE)	Nil
Mwesigye et al. (2013)	Numerical	Nil	Sytherm-800	Single phase	Turbulent flow Re number was in the range of 1.19×10^4–1.92×10^6	The entropy generation rate decreased when the inlet temperature increased. It is increased when the concentration ratio increased	Nil
Mwesigye et al. (2014a, b)	Numerical	Nil	Sytherm-800	Single phase	Turbulent flow Re number was in the range of 1.02×10^4–1.36×10^6	The total entropy generation increased as the rim angle and the fluid temperature decreased, and when the concentration ratio increased. The entropy generation rates were high at low rim angles because of high peak temperatures in the absorber tube	Nil

| Chang et al. (2014) | Experimental and numerical | Nil | Water | Single phase | Turbulent flow Re number was in the range of 1×10^4–3.5×10^4 | The wall temperature of an absorber tube varied with the circular angle (θ) of cross-section, and the maximum temperature occurred at $\theta =180°$ | The local temperature of the tube inside surface $T_{si}(\theta)$ can be calculated from the regression formula: $$T_{si}(\theta) = T_{si} - \Delta T_i \cos\theta = T_{f,ave} + \frac{q_{max}\, r_{s,o}}{0.023 Re^{0.8} Pr^{0.4} k_f}(1-\cos\theta)$$ The local temperature of the tube outside wall surface $T_{so}(\theta)$ can be calculated from: $$T_{so}(q) = \frac{\frac{1}{2} q_{max}\, r_{s,o} \ln\left(\frac{D_{s,o}}{D_{si}}\right)}{k_f}(1-\cos q) + T_{f,ave} + \frac{q_{max}\, r_{s,o}}{0.023 Re^{0.8} Pr^{0.4} k_f}(1-\cos q)$$ |

(continued)

Table 4.1 (continued)

References	Study type	Heat transfer enhancement technique	Fluid used	Phase mode	Flow type	Remarks	Proposed correlations
Wang et al. (2014)	Numerical	Nil	Molten salt	Single phase	Turbulent flow Velocity was in the range of 1–4 m/s	The CTD of the absorber increased with the rising of the DNI and decreased with the increase of fluid's inlet temperature and inlet velocity. The non-uniform distribution of the solar energy flux affected the CTD of receiver while it has less impact on the thermal efficiency	Nil
Sokhansefat et al. (2014)	Numerical/ CFD	Nil	Synthetic oil/ Al_2O_3 nanofluid	Single phase	Re number was in the range of 4000–18,000	The thermal enhancement due to the nanoparticles in the fluid reduced as the absorber operational temperature is increased	Nil
Ghasemi and Ahangar (2014)	Theoretical and numerical	Nil	Cu- water nanofluid	Single phase	Flow velocity was in the range of 0.5×10^{-3}–11.5 $\times 10^{-3}$ m/s	The nanofluid based linear parabolic concentrator has a higher efficiency than a similar conventional collector	Nil
Sunil et al. (2014)	Experimental	Nil	SiO_2 - water nanofluid	Single phase	Mass flow rate was in the range of 20–60 L/h	The results revealed that SiO_2-H_2O nanofluid has relatively higher thermal efficiency at higher Reynolds numbers	Nil
Wu et al. (2014)	Numerical	Nil	Therminol 55 and Gilotherm 55	Single phase	Turbulent flow Re number was 2×10^5	The temperature difference within the metal tube changed inversely with the fluid's velocity. The stagnation temperature of PTC receiver increased linearly with time, and it can reach at 700 K in about 130 s	Nil

Cheng et al. (2014)	Numerical	Nil	Thermal oil	Single phase	Not specified	The PTC of various configurational factors have different sensitivity levels to different optical errors. However, the optical accuracy requirements from various configurational factors of a PTC system were always unchanged	Nil
Mwesigye et al. (2015)	Numerical	Nil	Syltherm 800 - Al_2O_3 nanofluid	Single phase	Turbulent flow Re number was in the range of $3560-1.334 \times 10^6$	The thermal efficiency increased by up to 8% for the range of parameters considered. The utilisation of nanofluids became thermodynamically unpreferable beyond the optimum Reynolds number	Nil
Mwesigye et al. (2015)	Numerical	Nil	Syltherm oil - Al_2O_3 nanofluid	Single phase	Turbulent flow Re number is in the range of $3560-1.15 \times 10^6$	The PTC's thermal efficiency improved by up to 7.6% when using nanofluids. The entropy generated is a minimum at the optimal Reynolds number	$Nu = 0.008905 \ Re^{0.8966} \ Pr^{0.3805} \ \phi^{-1.1836 \times 10^{-3}}$ $f = 0.2085 \ Re^{-0.2132} \ \phi^{1.0538 \times 10^{-2}}$ $Re_{th} = 1{,}023{,}833 \ \phi^{0.02763} \Theta^{1.6259} Pr^{-0.7181}$ $Re_{opt} = 5{,}394{,}186 \ \phi^{1.2026 \times 10^{-3}} \Theta^{0.4883} \ Pr^{-1.0695}$

(continued)

Table 4.1 (continued)

References	Study type	Heat transfer enhancement technique	Fluid used	Phase mode	Flow type	Remarks	Proposed correlations
Mwesigye et al. (2015)	Numerical	Nil	Sytherm 800 - CuO nanofluid	Single phase	Turbulent flow Re number was in the range of 3.84×10^3– 1.35×10^6	The thermal performance increased up to 38% while the thermal efficiency increased up to 15%. High inlet temperatures and low flow rates provided remarkable enhancement in PTC's thermal efficiency	Nil
Wang et al. (2015)	Numerical	Nil	Downtherm-A synthetic oil	Single phase	Turbulent flow Velocity was in the range of 1–4 m/s	The CTD of the absorber decreased with the increases of inlet temperature and Reynolds number and it increased with the increment of the DNI. It was found that the absorber's thermal stress and deformations were higher than that of the glass cover	Nil
Zadeh et al. (2015)	Numerical	Nil	Al_2O_3-synthetic oil nanofluid	Single phase	Turbulent flow Re number was in the range of 4000–28,560	The thermal enhancement has a direct proportion with nanoparticle concentration while it has indirect proportion with the operational temperature	Nil

Tzivanidis et al. (2015)	Mathematical Model (1D)	Nil	Pressurised water (10–180 °C)	Single phase	Re number was in the range of 1000–9000	The efficiency of the collector was over 75% for high temperature levels. This was due to very low heat loss factor, which varied from 0.6 to 1.3 W/m²·K, depending on the inlet temperature	$\eta = 0.8 - 0.008843\left(\dfrac{T_{in} - T_{am}}{G_b}\right) - 0.00050558\,G_b\left(\dfrac{T_{in} - T_{am}}{G_b}\right)^2$ $U_L = 0.5222 + 1.8773\left(\dfrac{T_{in} - T_{am}}{G_b}\right)$ $K(\theta) = 1.0159 - 0.448\,\theta - 0.2985\,\theta^2$ $F_R = 0.9981\ \mathrm{F} =$ $0.9984\ \mathrm{F}' = 0.9997$
Li et al. (2015)	Numerical/ CFD	Nil	Superheated steam	Single phase	Re number was in the range of $2 \times 10^4 - 10^5$	It was not appropriate to utilise the experimental correlations for forced /combined convection to carry out the thermal design and prediction for PTC framework	Nil
Chaudhari et al. (2015)	Experimental	Nil	Al_2O_3 in water nanofluid	Single phase	Turbulent flow Re number was 5000	The PTC nanofluid-based has higher efficiency than PTC water-based. The solar thermal efficiency increased about 7% at 0.1% volume fraction. The nanofluid provided thermal enhancement by 32%	Nil

(continued)

Table 4.1 (continued)

References	Study type	Heat transfer enhancement technique	Fluid used	Phase mode	Flow type	Remarks	Proposed correlations
Basbous et al. (2015)	Numerical	Nil	Al_2O_3 in Syltherm 800 nanofluid	Single phase	Mass flow rate was 0.62 kg/s	The nanoparticles improved significantly the convection heat transfer by 18% and decreased the heat losses by 10%. The convection heat transfer factor tended to increase at high temperatures	Nil
Kasaeian et al. (2015)	Experimental	A black painted vacuumed steel tube, a copper bare tube, a glass enveloped non-evacuated copper tube and a vacuumed copper tube	Carbon nanotube -oil nanofluid	Single phase	Mass flow rate was not specified	The vacuumed tube has thermal efficiency 11% higher than the bare tube efficiency. The nanofluid showed high thermal potential for further and more complete examinations. A model of global efficiency was proposed and compared with the previous models	$\eta = 0.6730 - 0.2243 \dfrac{T_{f,i} - T_{amb}}{l}$
Basbous et al. (2015)	Numerical	Nil	Humid air, Syltherm 800	Single phase	Mass flow rate was 0.62 kg/s	The moisture variations increased the heat losses in the annulus in the case of low humidity and became closer to those of dry air for high humidity. The radiation heat losses in the annular space decreased by the moisture and the global thermal performances were slightly affected	Nil

Mwesigye et al., (2016a, b)	Numerical	Nil	Cu-Therminol VP-1 nanofluid	Single phase	Turbulent flow Flow rate was in the range of 1.22–1.35 m³/h	The thermal efficiency increased by 12.5% when the nanoparticle volume fraction increased from 0% to 6%	Nil
Bellos et al. (2016a, b)	Theoretical	Nil	Air, nitrogen, helium, neon, argon and CO$_2$	Single phase	Mass flow rate was in the range of 0.01–0.3 kg/s	Helium was the best working fluid for inlet temperature up to 700 K, while CO$_2$ was the most appropriate solution for higher temperature levels. The maximum exergetic efficiency achieved with helium at inlet temperature and mass flow rate of 640 K and 0.035 kg/s, respectively	Nil
Wang et al. (2016)	Numerical	Nil	Al$_2$O$_3$/synthetic oil nanofluid	Single phase	Flowrate was in the range of 2.8–3.3 L/min	The collector efficiencies Al$_2$O$_3$/ synthetic oil nanofluid were higher. The changes of temperature in the absorber with DNI, the inlet temperature and the inlet velocity were remarkably reduced	Nil
Ghasemi and Ranjbar (2016)	Numerical	Nil	CuO-water, Al$_2$O$_3$-water, nanofluids	Single phase	Turbulent flow Re number was in the range of 40,000–250,000	Nanofluid enhanced the PTC thermal performance compared with pure water. The heat transfer enhanced 28% and 35% with the utilisation of Al$_2$O$_3$-water and CuO-water nanofluids (ϕ = 3%), respectively	Nil

(continued)

Table 4.1 (continued)

References	Study type	Heat transfer enhancement technique	Fluid used	Phase mode	Flow type	Remarks	Proposed correlations
Abid et al. (2016)	Mathematical	Nil	Al_2O_3, Fe_2O_3, Molten salts (LiCl-RbCl/ $NaNO_3$-KNO_3)	Single phase	Mass flow rate was 0.01 kg/s	The overall exergy efficiency of PD and PT varied between 20.33% to 23.25% and 19.29% to 23.09%, respectively; with rise in ambient temperature from 275 K to 320 K. higher energetic and exergetic efficiencies was obtained when using nanofluids compared with molten salts for both cases	Nil
Li et al. (2016)	Numerical	Nil	Superheated steam with Pr = 1.5	Single phase	Laminar flow Re number was in the range from 250 to 1000	The free convective combined with forced convective increased the thermal rate by more than 10% when the Grashof number was greater than a threshold value. The combined flow and thermal features varied significantly with the solar elevation angle	$$\frac{Nu_{mixed}}{Nu_{forced}} = \left[1 + 0.0316\left(\frac{\phi+90}{180}\right)^{0.72295}\frac{Gr^{0.22834}}{Re^{-0.05491}}\right]^{1.28434}$$ $$\frac{f_{mixed}}{f_{forced}} = \left[1 + 0.01546\left(\frac{\phi+90}{180}\right)^{2.20042}\frac{Gr^{0.44621}}{Re^{-0.18795}}\right]^{-0.49873}$$ $$\frac{Nu_{mixed}}{Nu_{forced}} = \left[1 + 2.56\times10^{-5}\frac{Gr^{0.74884}}{Re^{-0.71355}}\right]^{-0.19574}$$ $$\frac{f_{mixed}}{f_{forced}} = \left[1 + 4.93808\times10^{-4}\frac{Gr^{0.39841}}{Re^{0.35179}}\right]^{-0.59787}$$

Reference							
Bellos et al. (2016a, b)	Numerical	Nil	CuO in Syltherm 800 and in nitrate molten salt (60% NaNO₃–40% KNO₃).	Single phase	Volumetric flow rate was in the range of 50–250 L/min	The thermal efficiency enhancement increased up to 0.76% when using oil-based nanofluids, while it increased up to 0.26% with the use of molten salt-based nanofluid	Nil
Kaloudis et al. (2016)	Numerical	Nil	Al₂O₃ in synthetic oil	Two phase	Re number was in the range of 4000–20,000	The nanoparticles enhanced the PTC's efficiency and it showed enhanced mixed convection effects for higher nanoparticle concentrations. A 10% boost in efficiency was possible for Al₂O₃ at concentration of 4%	Nil
Bortolato et al. (2016)	Experimental	Nil	Steam	Single phase	Mass flow rate was 250 kg/h	The optical efficiency was equal to 82%, while the thermal efficiency was around 64%, which was a promising value and can still be improved by using flat absorber and adopting a solar selective coating with a low thermal emittance. The time constant of the collector was equal to 213 s	Nil
Coccia et al. (2016)	Numerical	Nil	Fe₂O₃, SiO₂, TiO₂, ZnO, Al₂O₃ and au in water	Single phase	Mass flow rate is in the range of 0.5–1.5 kg/s	The nanoparticles of Au, TiO₂, ZnO and Al₂O₃ nanofluids at the lower concentrations, presented very slight improvements compared to the use of water. However, there was no advantage when increasing the concentration of nanoparticles	Nil

(continued)

Table 4.1 (continued)

References	Study type	Heat transfer enhancement technique	Fluid used	Phase mode	Flow type	Remarks	Proposed correlations
Ferraroa et al. (2016)	Numerical/ MATLAB	Nil	Synthetic oil–Al_2O_3 nanofluid	Single phase	Turbulent flow Mass flow rate was in the range of 1–7 kg/s	Only slight differences were observed for the power loss and the efficiency while the main advantage is represented by the lower pumping power	Nil
Basbous et al. (2016)	Numerical	Nil	Al_2O_3, Cu, CuO and Ag in Syltherm 800 nanofluid	Single phase	Turbulent flow Re number was in the range of 4000–40,000	The overall heat loss coefficient of the PTC was found to decrease when using nanofluids. The maximum thermal effectiveness was 36% using silver nanoparticles	Nil
Mwesigye et al. (2018)	Numerical	Nil	Copper-Therminol, silver-Therminol, Al_2O_3-Therminol nanofluids	Single phase	Turbulent flow Re number was in the range of 100–2,500,000	The thermal efficiency increased by 13.9%, 12.5% and 7.2% for Ag-Therminol, Cu-Therminol and Al_2O_3-Therminol, respectively, when the proportion ratio is 113	$(\eta_{th})_{opt} = 179\ (1 + \phi)^{0.04176}\ \theta^{-0.4166}\ C_R^{-0.1708}$
Bellos et al. (2017a, b, c, d, e)	Numerical	Nil	Supercritical CO_2	Single phase	Mass flow rate was in the range of 0.5–4 kg/s	Low-pressure levels (80 bar) had to be used in low temperatures while higher-pressure levels (200 bar) were proper for higher temperatures. The maximum exergetic efficiency was 45.3% for inlet temperature of 750 K	Nil

Ghasemi and Ranjbar (2017a, b)	Numerical	Nil	Al_2O_3-Therminol 66 nanofluid	Single phase	Turbulent flow Re number was in the range of 30,000–250,000	Nanofluid led to augment the thermal effectiveness. The Nusselt number increased by increasing the nanoparticle volume fraction	Nil
Huang et al. (2017)	Numerical	Nil	Therminol-VP1	Single phase	Turbulent flow Re number was 2×10^4 Gr number was in the range of 0–3.2×10^{10}	The thermal and flow fields factors in dimpled receiver tubes under NUHF were larger than those under UHF. The deep dimples (d/Di = 0.875) were far superior to the shallow dimples (d/Di = 0.125) at a same Grashof number	Nil
Bellos et al. (2017a, b, c, d, e)	Numerical	Nil	CuO- Syltherm oil 800, Al_2O_3- Syltherm oil 800	Single phase	Volumetric flow rate was in the range of 3–7 m³/h	The thermal effectiveness using nanofluids improved about 50% at higher temperatures. The thermal efficiency increased 1.26% and 1.13% with the utilisation of CuO and Al_2O_3, respectively, when the proportion ratio was maximised, and the flow rate was relatively low	Nil
Bellos et al. (2017a, b, c, d, e)	Theoretical using EES tool	Evacuated tube absorber	Pressurised water, Therminol VP-1, nitrate molten salt, sodium liquid, air, CO_2 and helium	Single phase	Mass flow rate was in the range of 0.1–2.5 kg/s	The maximum exergetic efficiency was 47.48% using liquid sodium at inlet temperature of 800 K, while it was 42.21%, 42.06% and 40.12% using helium, CO_2 and air, respectively. Pressurised water was the best working medium for temperature levels up to 550 K, while CO_2 and helium were only the good options for temperatures greater than 1100 K	Nil

(continued)

Table 4.1 (continued)

References	Study type	Heat transfer enhancement technique	Fluid used	Phase mode	Flow type	Remarks	Proposed correlations
Menbari et al. (2017)	Experimental	Nil	Al$_2$O$_3$, CuO in water nanofluids	Single phase	Mass flow rate was in the range of 10–100 L/h	The thermal efficiency of the PTC system enhanced by increasing nanoparticle volume fraction and nanofluid flow rate	Nil
Khakrah et al. (2017)	Numerical	Nil	Synthetic oil- Al$_2$O$_3$ nanofluid	Single phase	Turbulent flow Re number was in the range of 3150–21,000	The efficiencies reduced by 1–3% by rotating the reflector with 30° relative to wind direction. The addition of small amount of Al$_2$O$_3$ led to 14.3% and 12.4% efficiency enhancement, for horizontal and rotated reflectors, respectively	Nil
Kasaeian et al. (2017)	Numerical	A bare glass tube, non-evacuated glass-glass tube and a vacuumed absorber tube	Carbon nanotube, nanosilica in ethylene glycol nanofluids	Single phase	Mass flow rate was not specified	Carbon nanotubes have higher temperature and efficiency of 338.3 K and 74.9% in the vacuumed glass absorber tube. The carbon nanotubes have optimum point of volume fraction and thermal efficiency at 0.5, and 80.7%, whereas it is 0.4 and 70.9%, respectively, for nanosilica	Nil
Pavlovic et al. (2017a, b)	Numerical	Nil	Therminol VP-1	Single phase	Mass flow rate was in the range of 0.4–5 kg/s	Higher widths demanded higher receiver diameter for optimum performance. For inlet temperature equal to 200 °C, the optimum design was found to be 3000 mm width with 42.5 mm receiver diameter, with the focal length of 1840 mm	Nil

| Chafie et al. (2018) | Experimental | Nil | Transcal N thermal oil | Single phase | Flow rate was 0.2 kg/s | The useful energy gain transferred to the HTF and the exergy efficiency were proved to be smaller than the useful exergy rate and the exergy efficiency. The exergy factor heavily affected by the DNI and the operating temperature. The average energy efficiency, exergy efficiency and exergy rate for cloudy and sunny days varied between 19.72%, 8.51% and 0.08–36.1%, 11.72% and 0.12, respectively | Nil |
| Mwesigye et al. (2018) | Numerical | Nil | SWCNT-Therminol VP1 nanofluid | Single phase | Turbulent flow Re number was in the range of $1 \times 10^4 - 80 \times 10^4$ | The thermal efficiency increased around 4.4%, and the thermal effectiveness increased up to 234%, when the nanoparticle concentration increased from 0 to 2.5%. The reduction in entropy generation was about 70% | |

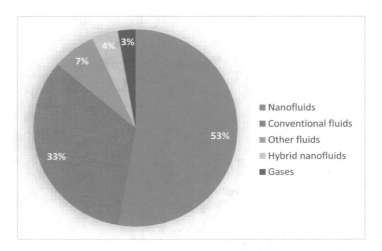

Fig. 4.42 Research efforts on PTC with different working fluids

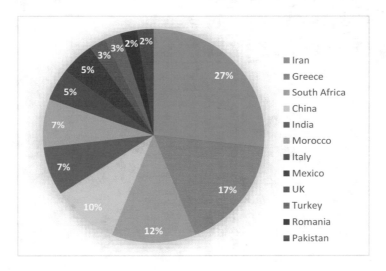

Fig. 4.43 Research efforts on PTC enhancement using nanofluids around the globe

12%, China with 10% and other countries with 2–7%. In contrary, with the usage of conventional fluids as PTC working medium, China has done great efforts in that with 42%, followed by Greece with 19%, then India with 10% and other countries have shared 3–7% as can be seen in Figs. 4.43 and 4.44.

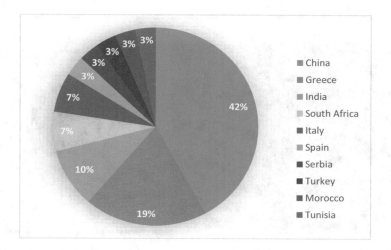

Fig. 4.44 Research efforts on PTC using conventional fluids around the globe

4.3 Hybrid Nanofluids

Recently, binary nanofluids terminology has been presented by suspending two kinds of nanoparticles in a base liquid (e.g., water, ethylene glycol or mixes of these two liquids). Binary nanofluids research began in 2013 and, their research advancement is still in progress. Uses of binary nanofluids in sustainable power sources, particularly in sun-oriented collectors are incredibly low referenced in the literature and there are extremely constrained specialised information accessible on the use of binary nanofluids in PTC. Subsequently, the utilisation of binary nanofluids acquire and more consideration. The binary nanofluid is a homogenous blend, which exhibits novel physical and chemical properties and distinctive thermal and hydraulic properties, that makes it amazingly a promising answer for better thermal qualities, contrasted to mono nanofluid because of synergetic impacts (Minea & El-Maghlany, 2018; Sidik et al., 2016; Babu et al., 2017).

In the literature, there are exceptionally uncommon examinations with binary nanofluids execution in sun-oriented collectors as can be seen beneath in the accompanying sections.

Menbari et al. (2017) designed and implemented an experimental apparatus to investigate the thermal effectiveness of binary nanofluids in direct absorption solar parabolic trough collectors (DASPTCs) and to evaluate the factors involved in their optimal stability as shown in Figs. 4.45 and 4.46. For this purpose, two dissimilar nanoparticles, i.e., CuO and c-Al$_2$O$_3$ were chosen to prepare a binary nanofluid. The outcomes demonstrated that the stability and thermal properties of binary nanofluids unequivocally rely upon pH, surfactant mass fraction and sonication time. It was noticed that the thermal characteristics and aggregation of the synthesised nanofluid were highest and lowest, respectively, under optimal stability conditions. The

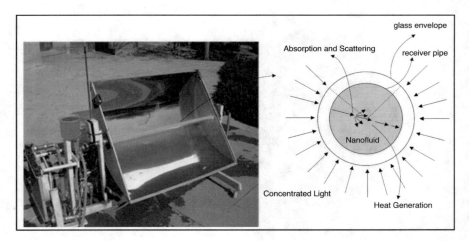

Fig. 4.45 The developed experimental system of the DAPTC collector Menbari et al. (2017). (License Number: 4620041052609)

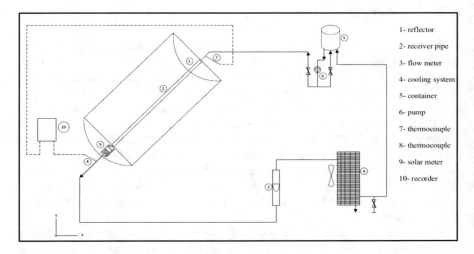

Fig. 4.46 Schematic of the hydraulic diagram of the DAPTC collector Menbari et al. (2017). (License Number: 4620041052609)

experiments revealed that the thermal effectiveness significantly enhanced with the rise of nanoparticle concentration and nanofluid mass flowrate. The results indicated that the thermal effectiveness of DAPTSC containing a mixture of two different nanoparticles scattered in water is more prominent than its partner scattered in a water-EG blend. This is mainly because EG can be progressively used for wide temperature range compared with water-EG blend which suffers from higher freezing and bubbling temperatures.

Bellos et al. (2018a, b, c) investigated the utilisation of single and binary nanofluids with Syltherm 800 in LS-2 PTC module by using a thermal analysis with the aid of EES as shown in Fig. 4.47a. The study covered various nanofluid types as 3% Al_2O_3/Oil, 3% TiO_2/Oil and 1.5% Al_2O_3–1.5% TiO_2/Oil, 300 to 650 K of inlet temperature and 150 L/min of mass flowrate. The thermal efficiency enhanced with 0.74% using binary nanofluid, while it was 0.341% and 0.340% using TiO_2 and Al_2O_3 nanofluids, respectively. The thermal efficiency enhanced up to 0.7% and 1.8% for single and binary nanofluids, respectively. The Nusselt number enhancement is found 121.7%, 23.8% and 23.4% for hybrid nanofluid, TiO_2 nanofluid and Al_2O_3 nanofluid, respectively. The binary nanofluid show to be more effective multiple times in improving the thermal effectiveness than the single nanofluids as appeared in Fig. 4.47b.

Minea and El-Maghlany (2018) reviewed the expansion of nanofluids research endeavours in solar frameworks application and to reveal the significance of utilising binary nanofluids in PTC as shown in Fig. 4.48. It was obviously shown an improvement in Nu number for all considered hybrid nanofluids. The highest increase in Nu number was noticed for 2% Cu-MgO binary nanofluid, and the thermal efficiency was maximum using 2% Ag-MgO-water binary nanofluid. The water-based binary nanofluid provides acceptable thermal augmentation with less

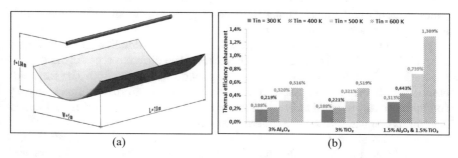

Fig. 4.47 (a) The LS-2 PTC examined module and (b) Thermal efficiency for various nanofluids Bellos et al. (2018a, b, c). (License Number: 4620041224384)

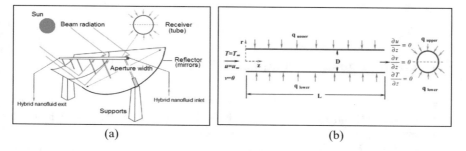

Fig. 4.48 (a) PTC system and (b) Schematic diagram of PTC receiver tube Minea and El-Maghlany (2018). (License Number: 4620041356313)

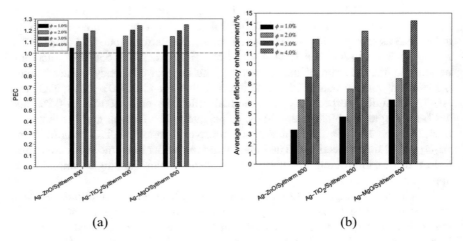

Fig. 4.49 (**a**) PEC variation and (**b**) thermal efficiency increment of PTC receiver tube for various binary fluids Ekiciler et al. (2020). (License Number: 4920041356414)

pressure loss penalty. Nevertheless, the addition of nanoparticles to 60EG:40 W is not recommended due to high-pressure drop penalty. It was inferred that binary nanofluids are excellent candidates for solar energy applications, even if the studies in the literature are limited at this moment.

Ekiciler et al. (2020) studied numerically 3D thermohydraulic features of binary nanofluids in a PTC's receiver tube. They used three different types of fluids, i.e., Ag-ZnO, Ag-TiO$_2$ and Ag-MgO and all dispersed in Syltherm 800. The Reynolds number was in the range of 10,000 and 80,000 using nonuniform heat flux boundary condition. It was found that the thermal and flow fields are superior for all binary nanofluids compared with the Syltherm 800. The PTC's thermal efficiency declined with Reynolds number increment and risen with the nanoparticle concentration. It was concluded that Ag-MgO/Syltherm 800 was highly effective fluid compared with other examined fluids at nanoparticle concentration of 4% as shown in Fig. 4.49.

A summary of literature studies on PTC utilising hybrid nanofluids are listed in Table 4.2.

4.4   Other Studies/Techniques on PTC

There are other few numerical research articles carried out by few researchers on thermal augmentation using various kinds of fluids and gases in PTC, which are comprehensively summarised below, in the following sections.

Bellos et al. (2016a, b) examined an energetic and exergetic comparison of various gases in a commercial PTC as shown in Fig. 4.50a using EES. Air, nitrogen,

Table 4.2 Summary of literature studies on PTC utilising binary nanofluids

References	Study type	Heat transfer enhancement technique	Fluid used	Phase mode	Flow Type	Remarks
Menbari et al. (2017)	Experimental	Nil	Al_2O_3, CuO in water nanofluids	Single phase	Mass flow rate was in the range of 10–100 L/h	The thermal efficiency enhanced with the nanoparticle volume fraction and nanofluid flow rate increment
Bellos et al. (2018a, b, c)	Numerical	Nil	3% Al_2O_3/ Syltherm 800, 3% TiO_2+ Syltherm 800 and 1.5% Al_2O_3–1.5% TiO_2 + Syltherm 800	Single phase	Volumetric flow rate was 150 L/min	The thermal effectiveness was up to 0.7% and 1.8% using single and binary nanofluids, respectively. The binary nanofluids provided 2.2 higher improvement in Nusselt number than single nanofluid using pure oil
Minea and El-Maghlany (2018)	Numerical	Nil	Cu-MgO in water, Ag-MgO in water, GO-CO_3O_4 in water, Al_2O_3-cu in water	Single phase	Re number was in the range of 100–2000	The 2% Cu-MgO binary nanofluid provided the highest increase in Nusselt number. The 2% Ag-MgO-water binary nanofluid exhibited the maximum efficiency
Ekiciler et al. (2020)	Numerical	Nil	Ag-ZnO/Syltherm 800, Ag-TiO_2/Syltherm 800 and Ag-MgO/Syltherm 800	Single phase	Re number was in the range of 10,000–80,000	The PTC's thermal efficiency declined with increasing Reynolds number and risen with the increasing nanoparticle concentration. It is found that Ag-MgO/Syltherm 800 nanofluid was highly effective fluid compared with other examined fluids

a

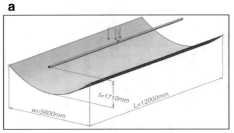

b

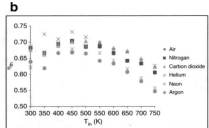

Fig. 4.50 (**a**) The examined LS-2 PTC module and (**b**) Thermal efficiency versus inlet temperature for various gases Bellos et al. (2016a, b). (License Number: 4620070074627)

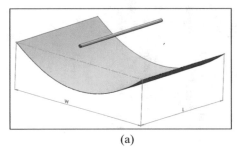

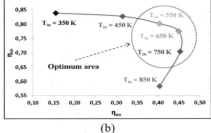

(a)   (b)

Fig. 4.51 (**a**) The examined module of LS-2 PTC and (**b**) Efficiency map of the PTC Bellos et al. (2017a, b, c, d, e). (License Number: 4620070282047)

carbon dioxide, helium, neon and argon were used. The results affirmed that helium, for inlet temperature up to 700 K, is the best working fluid, while CO_2 is the most appropriate solution for higher temperature levels. Air and nitrogen have similar performances, which are acceptable for utilisation in real systems. Argon is the lowest efficiency, and it is not suitable for operation in parabolic collectors. Neon performs well up to 380 K and beyond this value has low performance. The optimum exergetic efficiency achieved with helium operating to 640 K inlet temperature as shown in Fig. 4.50b. It was found that helium is the gas with the lower pressure drop (about 3 kPa) and for this reason has a high exergetic efficiency. It was proved that the use of alternative working fluids, as helium and carbon dioxide, could lead to higher exergetic output, compared with air, especially at higher temperature levels. This result establishes the use of these gases in many industrial applications and in power generation plants.

Bellos et al. (2017a, b, c, d, e) continued their research and investigated the utilisation of supercritical CO_2 in PTC as shown in Fig. 4.51a for various operating conditions. The results revealed that 0.5 kg/s is the optimum mass flowrate for maximising the exergetic effectiveness in low-temperature levels and 2.5 kg/s to be the optimum at higher temperatures. The exergetic analysis proved that low-pressure levels in the supercritical region (80 bar) are optimum for low temperatures (up to

550 K) and higher-pressure levels (200 bar) are suitable for temperatures levels over 650 K. The suggested efficiency map of the PTC is shown in Fig. 4.51b, which reveals that the optimum operating region is from 500 K to 800 K. The maximum exergetic efficiency is found 45.3% for inlet temperature 750 K, pressure 200 bar and mass flow rate 2.5 kg/s. The outcomes of this work could be utilised for the best possible design of sun-based power planted dependent on supercritical CO_2.

A summary of literature studies on PTC utilising gases are listed in Table 4.3.

Table 4.3 Summary of literature studies on PTC utilising gases

References	Study type	Heat transfer enhancement technique	Fluid used	Phase mode	Flow type	Remarks
Bellos et al. (2016a, b)	Theoretical	Nil	Air, nitrogen, helium, neon, argon and CO_2	Single phase	Mass flow rate in the range of 0.01–0.3 kg/s	Helium was the best working fluid at 700 K, while CO_2 was the most appropriate solution at higher temperatures. Helium working at 640 K inlet temperature and 0.035 kg/s mass flow rate provided the global maximum exergetic efficiency
Bellos et al. (2017a, b, c, d, e)	Numerical	Nil	Supercritical CO_2	Single phase	Mass flow rate is in the range of 0.5–4 kg/s	Low-pressure levels (80 bar) were suitable at low-temperature levels while higher-pressure levels (200 bar) were proper for higher temperature levels. The maximum exergetic efficiency was 45.3% for inlet temperature of 750 K

Chapter 5
Discussion on Heat Transfer Enhancement Methods

5.1 Introduction

In this chapter, PTC collector types have been tested for working with different types and shapes of passive techniques, and conventional fluids, nanofluids and hybrid nanofluids. Many nanofluids types have been used such as Al_2O_3, Cu, CuO, TiO_2, Fe_2O_3 and MWCNT. Most of the investigators have used nanofluids with water and oil as base fluids in their mathematical, experimental and numerical studies. High number of practical research used water-based nanofluids, with relatively small number of studies used thermal oils as base fluids. Then again, the numerical investigations incorporate both base fluids (water and oil). More importantly, many of the experimental and numerical studies have assumed the nanofluids as a homogenous mixture and a single-phase model is applied to simulate the nanofluids.

5.2 Discussion on Thermal Enhancement Methods

There are many studies on thermal analysis and computational tools for measuring PTC system performance used in industry. Either mathematical models, numerical models or experiments were used to evaluate the PTC thermal performances. Mathematical models use thermodynamics formulas to compute the thermal performance. Experiments are based on field measurements taken to obtain more realistic performance. The advantages of mathematical and numerical models are their simplicity and low cost when compared to experiments. These two models or groups are classified as shown in Fig. 5.1. The thermal performance of PTC using these methods are discussed in the next sections.

© The Author(s), under exclusive license to Springer Nature Switzerland AG 2023
H. A. Mohammed et al., *Parabolic Trough Solar Collectors*,
https://doi.org/10.1007/978-3-031-08701-1_5

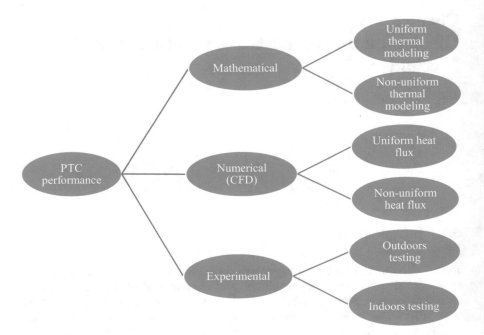

Fig. 5.1 Classification of thermal performance methods

5.2.1   *Mathematical Perspectives*

The general trend of the outcomes shows that the mathematical thermal models of PTC found a maximum thermal effectiveness is equal to 62.5%. The usage of nanofluids along with the thermal models has revealed that there is a small enhancement in heat gain (+0.3 W/m) and thermal efficiency (+0.03%). It is also noticed that the usage of nanofluids would only provide slight differences for the power loss and the thermal efficiency while the main advantage is lower the pumping power. It is advised that high working temperatures were increasingly reasonable for utilising nanofluids and produce higher relative additions of energy conveyed. The exergetic productivity improvement was higher priority than the energetic proficiency.

Additionally, the usage of conventional fluids numerically enhances the PTC's thermal effectiveness by more than 20–30%. There is a critical deviation in the discovered outcomes in view of the distinction in the followed approaches among the investigations. In this way, now, it is fascinating to examine and condense the major results for each technique utilised. It is revealed that the conventional fluids (therminol oil or Syltherm oil) have better PTC thermal effectiveness and a lower pressure loss than other molten salt fluids. Besides that, the entropy generation rate decrease was increased when the fluid inlet temperature and the concentration ratio increased and when the rim angle decreased. The usage of molten salt in PTC increased the CTD of the absorber with DNI rising and decreased when higher inlet temperature and inlet velocity used. The numerical outcomes indicate that the non-uniform solar flux significantly affects the CTD of absorber while it has a small impact on the thermal effectiveness.

It was seen that the absorber tube's material plays a critical factor on the PTC structural effectiveness. It is discovered that the adjustment in absorber tube's material, e.g., steel and copper, aluminium, bimetallic and tetralayered negligibly affects the heat transferred to the HTF, yet it significantly affects the twisting because of thermal expansion just as because of self-weight. It is found that a tetra-layered which had a lower weight and improved temperature profile and provides a reduction in the optimum deflection by 45–49% when contrasted with steel.

5.2.2   Numerical Perspectives

Interestingly the numerical results found that it is not appropriate to utilise the experimental correlations for forced convection or traditional mixed convention to carry out thermal design and prediction for PTC. Additionally, the free convective combined with laminar forced convective increased the thermal effectiveness by more than 10% when the Grashof number is greater than a threshold value. The combined flow and thermal features vary with the solar elevation angle where the thermal deterioration occurs at Richardson number greater than 12.8 (Li et al., 2015). It was found numerically that the receiver works very well when the fluid's inlet temperature was less than 150 °C (with an efficiency sacrifice within 4%), and particularly in the range of 100–120 °C (with an efficiency sacrifice within 1.5%). Several strategies were suggested to be effective including using non-selective coating material instead of the selective one for the absorber, reducing the outer fluid's velocity and lowering the emissivity of the inner glass tube in specific conditions (Fan et al., 2018).

Furthermore, the usage of nanofluids in numerical studies dramatically augment the PTC's thermal effectiveness. There was a critical deviation in the enhancement rate due to the distinction in the followed approaches among the investigations utilising CFD, FVM, FEM, EES and OpenFoam. Thus, it is important to highlight and summarise the main results for every method used. Most of the numerical outcomes of a PTC receiver with the use of nanofluids exhibit high thermal and thermodynamic effectiveness by up to 8% compared with a similar conventional-based fluid collector. It is pointed out using CFD models that the heat transfer coefficients present an ascending pattern with Reynolds number increment. The heat transfer improvement has an immediate association with the nanoparticle volume fraction, although it has an opposite association with the operational temperature. The results further portray that there is a specific Reynolds number value past which the utilisation of nanofluids turns out to be thermodynamically bothersome at a given inlet temperature.

Generally, the enhancement percentage varied for different researchers as they used different operating and boundary conditions in their studies. For instance, the thermal effectiveness and the thermal efficiency increased up to 38% and 15%, respectively, as indicated by Mwesigye et al. (2015). The nanoparticles improve significantly the convection coefficient by about 18% and could decrease the heat losses of about 10% (Basbous et al., 2015). However, the thermal efficiency increased by 12.5% and the entropy generation rates reduced as the nanoparticle concentration changed from 0% to 6% (Mwesigye et al., 2016a, b). Moreover, the changes of the maximum temperature in the absorber with DNI, inlet temperature and inlet velocity are remarkably reduced (Wang et al., 2016). The utilisation of Al_2O_3-water and CuO-water nanofluids at $\phi = 3\%$ provided enhancement in heat transfer up to 28% and 35%, respectively, as observed by Ghasemi and Ranjbar (2016). Moreover, the utilisation of oil-based nanofluids and molten salt-based nanofluid boosted the thermal efficiency up to 0.76%, and 0.26%, respectively. The utilisation of Syltherm 800-CuO and molten salt-CuO boosted the Nusselt number up to 40% and 13%, respectively, according to Bellos et al. (2016a, b). It was noticed that Au, TiO_2, ZnO and Al_2O_3 nanofluids exhibited minimal improvements at lower concentrations, contrasted to the utilisation of water, while there was no advantage with respect to water when the concentration of nanoparticles increased (Coccia et al., 2016). The maximum augmentation in PTC's thermal effectiveness was recorded for silver nanoparticles which reach 36% rise in heat transfer and 21% decrease in overall heat loss coefficient (Basbous et al., 2016). Other types of nanofluids such as silver-Therminol, copper-Therminol and Al_2O_3-Therminol show increase in the thermal efficiency by 13.9%, 12.5% and 7.2%, respectively, when the proportion ratio is 113 (Mwesigye et al., 2018). In contrast, the thermal enhancement is proved approximately to be 50% utilising nanofluids and it is greater at high temperature values. The thermal efficiency increased up to 1.26% and 1.13% with the use of CuO and Al_2O_3, respectively, when the concentration ratio is maximised and the flow rate is relatively low (Bellos et al., 2017a, b, c). Moreover, the efficiency enhancement for horizontal and rotated reflectors was 14.3% and 12.4%, respectively, when 5% of Al_2O_3 was added to the base synthetic oil (Khakrah et al., 2017).

Furthermore, the usage of carbon nanotubes in the vacuumed glass absorber tube increased the efficiency and temperature to 74.9% and 338.3 K, respectively. The carbon nanotubes exhibited optimum values of volume fraction and thermal efficiency at 0.5 and 80.7%, respectively, whereas for nanosilica it is 0.4 and 70.9%, respectively (Kasaeian et al., 2017). The usage of SWCNTs significantly improved the thermal effectiveness with little rise in the thermal efficiency, while the thermal effectiveness, thermal efficiency boosted up to 234% and 4.4%, respectively, with entropy generation decrease of 70% using volume fraction of 2.5% (Mwesigye et al., 2018). Furthermore, the utilisation of nanofluids with oil-based and molten salt-based provides thermal efficiency boost up to 0.76% and 0.26%, respectively (Bellos al., 2018a, b, c). Moreover, the increase in thermal effectiveness with CuO nanofluid is more than with Al_2O_3 and SiC nanoparticles (Marefati et al., 2018). However, the CuO nanoparticle exhibited a peak exergy efficiency of 9.05% and the maximum daily gain of thermal energy delivered was 1.46% with the use of 5% Al_2O_3 (Allouhia et al., 2018).

5.2.3   Experimental Perspectives

From the experimental perspectives, the outcomes revealed that there is a need of extra experimental investigations on the usage of conventional fluids and nanofluids in PTC systems. However, these experimental studies have also confirmed the PTC's thermal effectiveness boost. These studies have also confirmed that the PTC's thermal effectiveness relies on various factors. For instance, the experimental optical efficiency is equal to 82%, while the thermal efficiency is around 64%, and it can still be improved by using a flat absorber as a piece of a cavity receiver and by covering the absorber tube with a low thermal emittance based on (Bortolato et al., 2016). In addition, the energy and exergy efficiencies along with the exergy rate varied, for the cloudy and sunny days, between 19.72%, 8.51% and 0.08 to 36.1%, 11.72% and 0.12, respectively (Chafie et al., 2018). The PTC's efficiency diminishes when the working fluid exit temperature increases (Kumar and Kumar 2018).

Furthermore, the experimental studies using nanofluids have also shown higher enhancement of PTC systems contrasted to classical fluids. The experimental results have revealed that the PTC's thermal efficiency could be enhanced by using higher nanoparticle volume fraction and nanofluid flow rate. The PTC's thermal efficiency of nanofluids strongly replies on the incident angle, where it is maximum at the lowest angle. For instance, the nanofluid causes the solar thermal efficiency increased by 7% and the thermal factor enhancement by 32% (Chaudhari et al., 2015). This is also supported by the results of Kasaeian et al. (2015) which show that the vacuumed tube thermal efficiency is 11% higher than the bare tube efficiency. This is also further confirmed by Subramani et al. (2018) results as the maximum thermal efficiency boost using TiO_2 nanoparticle was 8.66% higher than water. The energy factor was 9.5% higher than that of water. In addition, the maximum efficiencies were 13% and 11% higher using Al_2O_3 and Fe_2O_3 nanoparticles, respectively, compared to water under similar working situations (Rehan et al., 2018). The maximum thermal efficiencies varied between 52.4–57.7% using 3% of nanoparticles and 40.8–46.5% using water (De los Rios et al., 2018). The PTC's efficiency increased to more than 25% with the existence of external magnetic field, than the classical PTC (Alsaady et al., 2018).

It can be observed from the above results that the PTCs based nanofluid spread the most extensive piece of the literature. There was an extraordinary assortment of experimental and numerical examinations utilised diverse nanofluids and different ordinary fluids. The experimental studies demonstrate that there are significant improvements of about 11% with mean thermal proficiency. The numerical examinations give lower values of thermal improvements contrasted with the experimental contemplates. More explicitly, the CFD investigations demonstrate a thermal efficiency improvement of 7%, whereas the thermal models give an improvement of 1%. These various qualities demonstrate that the CFD and the thermal models utilise various strategies/suppositions, which are related to nanofluids. This urges the requirement for multilateral investigations for PTC frameworks, which incorporate both practical and theoretical examinations at the same operating conditions so comparable results could be obtained with high accuracy.

It is worth noting to mention that there is only one numerical study done by Kaloudis et al. (2016) on modelling the nanoparticles in two-phase flow in PTC framework. It is noticed that the existence of nanoparticles also enhanced the PTC's overall thermal effectiveness. With 10% boost in the thermal efficiency using 4% of Al_2O_3 nanoparticle. Thus, there is also a high need for more experimental and numerical investigations to be carried out for PTC systems at the same operating conditions using two-phase models.

It is also worth mentioning that there are only few numerical studies performed on the usage of hybrid nanofluids in PTC. These studies revealed that the binary nanofluid and single nanofluid produced up to 1.8% and 0.7% in thermal efficiency boost, respectively (Bellos et al., 2018a, b, c). They also stated that the PTC's efficiency increased while Re number increased and the maximum efficiency was obtained using 2% of Ag-MgO-water binary nanofluid (Minea & El-Maghlany, 2018). Thus, there is also a high urge for more experimental and numerical investigations to be carried out using various types of hybrid nanofluids to examine their efficacy on the PTC thermal, hydraulic and exergetic performance.

5.2.4   Passive Techniques Perspectives

Another aspect of this chapter is the usage of passive techniques for improving the PTC system performance. The numerical studies of PTC using passive techniques cover most extensive piece of the available literature because of the recent development in computers, programming languages and experimental techniques. It thus makes it easier for the researchers to conduct various computational studies under different geometrical and operational conditions. There is an incredible assortment of numerical studies using conventional fluids and nanofluids associated with different types of passive techniques. Most of the numerical results show improved PTC's thermal effectiveness with the use of passive technique by different percentage compared with a plain collector. They also stated that the passive techniques greatly reduced the Q_{loss}, T_{max} and circumferential temperature difference. In fact, various passive technique geometries and different fluids used by various researchers are the main reasons for the changes in thermal efficiency and performance. Many researchers have used the twisted tape insert as it can remarkably enhance the consistency of PTC tube wall's temperature profile. It can also enhance the heat transfer effectively with increasing the friction factor especially when using tight-fit twisted tape. Researchers have also found that the geometrical parameters of the passive technique such as length, thickness, width, depth, distance, have a remarkable influence on the PTC's thermal and hydraulic effectiveness. The advantages of passive technique are to unify the temperature profile and to reduce the maximum temperature on the absorber's tube under certain operational conditions (flow and geometrical parameters, etc.) which consequently diminish the thermal deformation and boost the endurance of the PTC receiver.

For instance, the consideration of permeable inserts in a PTC receiver using Therminol-VP improved the thermal effectiveness around 17.5% with a pressure loss of 2 kPa (Reddy et al., 2008; Kumar & Reddy, 2009, 2012). There was a 3% improvement of the collector efficiency using helical internal fins with Syltherm 800 oil. This outcome proposed an improvement of 2%, with a decline in the maintenance cost (Munoz & Abanades, 2011). The thermal loss of the PTC using unilateral longitudinal vortex generators with oil syltherm-800 reduces by 1.35–12.1% to that of the smooth PTC. The PTC has greater overall thermal effectiveness than that of the smooth PTC (Cheng et al., 2012a, b). The PTC pipe without helical internal fins suffers from higher temperature gradient (around 20 K) compared with the pipes with helical fins (Aldali et al., 2013). The maximum CTD on the external surface of receiver's pipe decreased by 45%, which extraordinarily reduced the thermal deformation (Wang et al., 2013). The PTC's thermal efficiency was remarkably improved using porous two segmental rings (Ghasemi et al., 2013), perforated louvered twisted tape (LTT) (Ghadirijafarbeigloo et al., 2014), dimples receiver contrasted to that with protrusions or fins (Huang et al., 2015) and the introduction of wire coils (Diwan & Soni, 2015). The modified thermal efficiency increased between 1.2% to 8% for a PTC equipped with perforated plate insert using sytherm-800 (Mwesigye et al., 2014a, b). It is also significantly improved with the use of nanofluid (Al_2O_3-synthetic oil) up to 15% (Mwesigye et al., 2015). There was considerable increase, using wall-detached twisted tape inserts, in thermal efficiency up to 10%, thermal effectiveness of 169%, decline in CTD to 68% and entropy generation rate reduction of about 58% compared with a plain absorber tube (Mwesigye et al., 2016a, b). The usage of asymmetric outward convex corrugated tube as receiver can provide thermal effectiveness of 148% and maximum thermal deformation of 26.8% (Fuqiang et al., 2016a, b). The concentric and eccentric pipe inserts with molten salt can remarkably enhance the thermal effectiveness around 1.11 ~ 1.64 than a PTC without inserts (Chang et al., 2017).

Furthermore, the PTC absorber tube equipped with internal fins boosted the thermal efficiency and thermal effectiveness factor by 1.27% and 1.483, respectively (Bellos et al., 2017a, b, c, d, e). The PTC receiver tube equipped with pin fin arrays inserting increased the Nusselt number and the thermal effectiveness factor by 9% and 12%, respectively (Xiangtao et al., 2017). Very recently, the PTC receiver tube equipped with flow inserts having star shape boosted the thermal, exergy and overall efficiencies by 1% (Bellos et al., 2018a, b, c). The utilisation of internal rectangular finned tube fins alone and CuO-oil nanofluid alone provided 1.1% and 0.76% improvement in thermal effectiveness and the combination of both techniques provided 1.54% improvement (Bellos et al., 2018a, b, c) The usage of S-curved sinusoidal tube receiver with synthetic oil increased the maximum performance about 135%. The converging-diverging PTC absorber tube provided the highest exergetic improvement of 0.65% and 0.73% using oil and Al_2O_3-oil nanofluid, respectively, compared with other absorber geometries (longitudinal finned absorber, twisted tape tube) (Okonkwo et al., 2018). The oblique delta-winglet twisted tape insert increased the thermal and exergy efficiencies by 12.05% and 4.95%, respectively. The serrated twisted tape inserts, among all other insert types (square cut, oblique delta-winglet, alternate clockwise and counter-clockwise), increased the thermal and exergy efficiencies by 13.63% and 15.40%, respectively, under the same conditions (Rawani et al., 2018a, b).

It is interesting to highlight in this chapter that diverse scientists in various flow characteristics (i.e., laminar and turbulent flows) also experimentally investigated the thermohydraulic performance of PTC. The PTC performance was found different for both flows, as it was more compelling than laminar flow and it is associated with high-pressure loss. Apart from cross-section configuration, investigations were conducted on twisted tape inserts, coiled insert, fins, ribs, corrugated tube, porous disc, metal foam, and high augmentation was found with these techniques due to high surface to volume ratio. The previous studies also portrayed that high thermal effectiveness is related with these enhancement techniques. The production of swirl stream utilising inserts tapes with different geometries is helpful to increase the Nusselt number. The thermohydraulic performance also relies on the passive technique's geometry. A lot of adjustment on straightforward and sophisticated tapes was made by researchers and discovered enhancements in their outcomes. However, the experimental studies using passive techniques-based PTCs were little compared with the numerical studies mentioned above because of the cost associated with constructing PTC systems with its instrumentations. There is a variety of these studies using conventional fluids and nanofluids. Most of these studies have stated that, for example, the enhancement efficiency is 135–205% over plain absorber/receiver of PTC with twisted tape inserts utilising silver-water nanofluid (Waghole et al., 2014). The highest efficiency of the coiled solar receiver using air is 82% and the highest exergy efficiency is approximately 28% (Zhu et al., 2015).

Moreover, the thermal effectiveness can be increased by using an absorber having internal multiple-fin array longitudinally welded, covered with selective oxide layer and mounted inside a single glass pipe. This enabled to obtain a vacuum from the glass cover which reduces the heal loss (Nems & Kasperski, 2016). The absorber filled with metal foam has a positive effect on the collector efficiency due to thermal conductivity enhancement compared with applying copper foam inside absorber (Jamal-Abad et al., 2017). The water was the most appropriate working fluid compared with air, and Therminol VP-1, in corrugated spiral cavity receiver as it can proficiently work at lower temperatures, while the thermal oil is the best at higher temperatures. The exergetic examination indicated that the air and thermal oil are the best decision in low and high temperature values, respectively (Pavlovic et al., 2017a, b).

In addition, the optimum value of various passive technique parameters such as twist angle, depth of wing cut ratio, space ratio, depth ratio, width ratio, pitch ratio, etc. are also evaluated in each study. It can be observed that for different studies there was different overall enhancement ratio with different factors studied. Experimentally, receivers equipped with twisted tape inserts geometries show high thermal performance compared with coiled, porous disc, internal fins, metal foam and corrugated spiral receivers. It can likewise be noticed that the total improvement is much better at higher Reynolds number values which is associated with higher pressure drop.

It can also be observed that there was 2–3% improvement in the useful output (heating, cooling or electricity) and these outcomes demonstrate that the utilisation of nanofluids can be used successfully in genuine PTC frameworks. Nonetheless, this moderately little improvement can increase the monetary status of the framework and build its yearly yield with good annual revenue. Additionally, the utilisation of nanofluids can diminish the heat exchange surface regions and consequently decreases the collecting region and minimises the land usage to install the PTC frameworks. Besides, the utilisation of nanofluids can diminish the CO_2 discharges in the process of heat generation or power generation.

Another important aspect of this chapter is the usage of various kinds of gases in numerical studies of PTC systems. It also improved the exergetic performance of the PTC systems by high amount. There is a variation in the found outcomes because of the difference in the operational conditions used in every study. For instance, the optimum exergetic efficiency is 45.3% achieved with helium. However, it is 25.62% for operation with air, while for therminol VP-1 is 31.67% for 500 K and 100 L/min (Bellos et al., 2016a, b). It can thus be seen from this very little literature of PTC system operated with gases that there is a need for further experimental research to be conducted to check the applicability of using different types of gases in PTC systems.

It can also be noticed for PTC-based binary nanofluids that there are just few experimental studies, and thermal effectiveness improvement of 13% was observed. It is clear that there is a requirement for increasingly experimental studies for PTC utilising various kinds of binary nanofluids. The outcomes demonstrate that the exergetic augmentation with the utilisation of binary nanofluids is significant and thus the utilisation of PTC-based binary nanofluid for power generation is suggested and promising.

It can also be inferred that the high thermal and exergetic efficiency improvements in the practical and theoretical studies demonstrate that the nanofluids and binary nanofluids are fluids of the future. In any case, there is a requirement for building up measures for performing appropriate investigations with nanofluids. Also, the specialists need to consider that the frameworks must be tried under the equivalent working conditions and to give high accentuation in the best possible assurance of the optical efficiency for each situation. Additionally, the thermophysical properties of each fluid used need to be experimentally quantified and not just depending on theoretical equations that are available in the open literature to estimate these properties.

5.3   Challenges and Research Gaps

5.3.1   Challenges

New regions with very energising potential applications have opened utilising nano-fluids and binary nanofluids. Nevertheless, there are a couple of primary concerns, which must be contemplated later on in future investigations. To build up the utilisa-tion of nanofluids in PTC frameworks, the cost of nanofluid planning and union ought to be considered. On the off chance that the exchange offs between the addi-tional expense and execution enhancements are not all around affirmed, at that point the utilisations of nanofluids may be restricted. In this way, it is significant to con-template both nanofluids cost examination and execution streamlining simultane-ously. Nevertheless, such examinations considering these investigations are rare. Subsequently, future investigations ought to consider the financial examination alongside the expenses brought about for nanofluids generation.

It could be seen from the outcomes of this book that nanofluids offer a few allur-ing and helpful properties that can enhance the exhibition of PTC frameworks. Specifically, thermal conductivity improvement found in various nanofluids has urged numerous scientists to assess their performance in PTC frameworks as found in Table 4.1 (Chap. 4). The consequences of a few examinations are displayed in Table 4.1 (Chap. 4), which can affirm that the expansion of little amounts of nanoparticles can remarkably enhance the properties of working liquids. In any case, investigations have additionally recognised various components (i.e., mole-cule bunching, agglomeration and precipitation) that block the long-term steadi-ness, the possibility of the nanofluids in genuine applications and suitability of nanofluids for pragmatic use in sun-oriented thermal applications.

The long-term stability of nanoparticles suspensions is one of the significant issues for both applied and practical investigations. Planning strategy is a funda-mental job in the final nanofluid product quality. It is commonly to add surfactants or added substances to the nanofluid to influence its surface science, diminish the surface tension occurs in the liquid, enhance the nanoparticle suspension in a liquid and it would eventually augment the overall properties of the nanofluid. For instance, expanding convergences of surfactants will increase nanofluid consistency and pro-duce larger pressure loss through the PTC framework. This will cause higher pump-ing power to circulate the nanofluid fluid, which is deemed a significant issue and must be investigated in detail. Subsequently, there is a requirement for appropriate computation of the extra pumping power. Thus, it is suggested that the correct crite-ria to assess both the thermal augmentation and the higher pumping work are made using the exergy and the general effectiveness.

Besides, factors, for example, evolving thermo-physical properties with temper-ature, molecule size variety, changing molecule volume division and photograph thermal debasement all require further investigations. For instance, expanding the volume division of added substances will expand the nanofluids density and viscos-ity, which will have higher-pressure loss through the PTC framework, which requires a higher pumping power to flow the working liquid. A regular measurement of surfactant amount can significantly influence the consistency and stability as well as the thermal properties of the nanofluids. Thus, obtaining stable nanofluids with high-volume fraction is the most challenging steps for future investigation.

Another challenge associated with the increment of surfactants amount in the preparation process of nanofluids at high temperature levels as the added surfactants may be decomposed causing detrimental effects. It is imperative to evaluate the nanofluid quality before its applied in any application and this is dependent on the used technique rules to standardise the experimental outcomes from various research groups. The standardised experimental procedure will provide simpler and steady evaluation of the experimental outcomes among various groups without the altering nanofluid characteristics that may influence the final outcomes of the research. Along these lines, top to bottom hypothetical and experimental examinations are expected to help the exact and systematic comprehension of parameters affecting the nanofluids performance. Additionally, there is an irregularity among the outcomes announced by various research groups. The need to a detailed clarification to the errors associated with the experimental outcomes is necessary along with providing exact combination and portrayal of nanofluids following a standardised methodology. There is also a need to look more into nanofluid's material science to analyse some confusions about the controlling parameters responsible for heat transfer enhancement.

5.3.2   Research Gaps

It can be inferred from the outcomes that the majority of investigations until date are mostly centred on nanofluids with water as base fluid than oil. Therefore, extra investigations on nanofluids with oil is fundamental to examine their impact on PTC frameworks carefully. More consumption tests are essential to evaluate the impacts of long-term nanofluid runs in PTC frameworks. The nanofluids stability ought to be examined for different PTC frameworks utilising, for example, functionalised nanoparticles. Another important issue is that there is a need for more research on the usage of hybrid nanofluids as well as new types of gases such as binary gas mixtures in PTC systems to check their efficacy in such systems.

Most of the investigations have centred their attention on PTC systems either with or without passive techniques, whereas very little investigations are reported showing PTC systems using hybrid nanofluids. Mono and binary nanoparticles can be utilised with aqueous and oil as base fluids to examine PTC systems performance. Additionally, environmental impact, exergetic and life cycle analyses should be done to evaluate the reasonability of utilising binary nanofluids in PTC frameworks and other solar systems.

There are some research gaps that can be highlighted from the present outcomes and future directions for additional research are recommended below.

- The development of new heat transfer fluids needs to be intensified and a consistent theory is urgently required to execute new thermal fluids in real solar energy applications.
- Another important topic to be addressed is the report of the thermophysical properties of hybrid nanofluids, as well as to propose solid correlations to estimate the properties of these new heat transfer fluids.
- In solar energy applications, fluid flow is generally in the laminar regime. Therefore, experimental as well as numerical endeavours are considerably required to fully understand the behaviour of hybrid nanofluids in laminar, transition and turbulent flows.
- The development of new PTC using passive technique with different shapes such as wavy, zigzag, corrugated tube with and without inserts combined with hybrid nanofluids also needs to be conducted to distinguish their effects on the PTC overall performance.
- There is a need to experimentally test different types of gases in PTC system using passive technique with different shapes and different temperature ranges to check their suitability in enhancing the PTC system overall performance.

It is also significant to state that there are strong future trends of using nanofluid or hybrid nanofluids in PTC frameworks. As it has been elaborated in the above previous sections, there are new brainstorming about implementing binary nanofluids with magnetic fields in PTC frameworks and it needs to be thoroughly investigated. In addition, the coupling of these different advanced fluids with various passive techniques (inserts, modified tubes, different shapes, etc.) is an encouraging decision as indicated by the above literature results. Moreover, there is a requirement for progressively experimental contemplates to utilise nanofluid/binary nanofluids based PTC frameworks in enormous scale establishments. The performance upgrades of the incredible sunlight-based field will demonstrate the genuine improvement, and this will be a basic advance for the foundation and commercialisation of nanofluid/binary nanofluids based PTC frameworks.

Chapter 6
Conclusions and Recommendations

6.1 Conclusions

This book surveys the approaches used (mathematical, experimental and numerical) on thermal augmentation techniques of PTC systems. There are levels of popularity of giving higher thermal rate and lower pressure loss in PTC frameworks. The utilisation of modified surfaces such as discs, fins, wire coil, twisted tape or any surface modifications can augment the heat transfer with its surface area as it causes high turbulence intensity and strong secondary flows. Moreover, there important parameters need to be considered to improve the thermal effectiveness of PTC frameworks and one of them is the choice of the working fluids type with various operating temperatures. Another vital aspect is the thermal energetic and exergetic analyses to identify the thermal and exergetic efficiencies of PTC frameworks. In this review, the thermal effectiveness augmentation with conventional fluids, single nanofluids and binary nanofluids as well as gases additionally featured, along with the nanofluids benefits are obviously portrayed. Additionally, the difficulties about the utilisation of these fluids in this solar research area are likewise examined. The major findings from this current article are outlined below:

- The utilisation of inserts gives comparable outcomes in the thermal effectiveness augmentation of the PTC framework. Notwithstanding, this strategy presents points of interest contrasted with the utilisation of inner fins or nanofluids as it tends to be applied in the current PTC without alterations which have many working difficulties.
- The use of fins with high number produces to larger effectiveness; however, it is recommended that the location of fins must be in PTC's lower part. The fins located at the PTC's upper part did not provide remarkable improvement in the PTC performance.
- The wavy-tape insert is more viable and appropriate to be utilised more effective and applicable in PTC when the system is operated at a relatively lower flow rate

as the merits of wavy PTC slowly diminished as the fluid mass flow rate increased due to the high increment of flow resistance.

- Utilising nanofluid in PTC has several advantages and causes augmenting the thermal field, thusly diminishing the size of cylinders and heat exchangers. Consequently, they diminish the thermal stress and the receiver's deformation. However, these benefits can be obtained with some modifications (e.g., materials, framework design and capital cost) on the PTC framework.
- The results revealed that with expanded nanofluid volume fraction proportions, receiver thermal execution debases, for both PTC heat loss and CTD expanding, particularly at low stream rates.
- The metallic (Cu, SiC) nanoparticles enhance heat transfer incredibly than different nanoparticles (C, Al_2O_3). It is found that combining both mechanisms (fins and nanofluid) provides excellent heat transfer and greater thermo-hydraulic performance.
- The PTC investigations using nanofluids are the majority in the available literature. The experimental examinations demonstrate significant enhancement near 11–13%, while the theoretical investigations provide lower value. The CFD models produce thermal effectiveness of 8%, while the thermal analyses provide improvements of 1–2%.
- The copper is a suitable option to be utilised as PTC's absorber pipe material in terms of thermal strain and deformation.
- The results clearly indicated that the nanofluids have higher thermal effectiveness than classical fluids. These augmentations are responsively small because of the thermal losses associated with PTCs; however, these augmentations can make the PTCs an increasingly feasible and good energy innovation.
- It is proved that the use of alternative working fluids, as helium and carbon dioxide, can lead to higher exergetic output, compared with air, especially in higher temperature levels. This result establishes the use of these gases in many industrial applications and in power generation plants.

6.2   Recommendations for Future Work

The following suggestions are presented for further research:

- Few investigators have tested the PTC using nanofluids in real power generation systems. The outcomes indicated 4% enhancements in system performance. However, there is a need for a more detailed experimental investigation on real applications.
- Further investigations on the whole photo-thermal process of the PTC system under real operating conditions over a year are needed to be further performed. This could be carried out by combining the absorbed nonuniform solar radiation distributions calculated by numerical models with an ensuing thermal analysis as heat sources.

- The practicality of using nanofluids in extraordinary scale PTC applications should be inspected more in-depth experimentally to find solutions of nanofluids issues such as its stability, agglomeration, arrangement, etc.
- The utilisation of nanofluids can make an increasingly minimised framework and to decrease the land usage. Moreover, the CO_2 emissions can be diminished in view of the expanded effectiveness. In this way, the examination right now is pivotal, since the integration of nanofluids in PTC frameworks can convey another option, eco-accommodating and feasible sustainable power source.
- Moreover, the expanded pumping work request must be considered regardless in any situation. The performance assessment paradigm and the exergy productivity are the most common criteria, aside from the thermal effectiveness.
- There is a need to prove which fluid is the most proper for every temperature level of PTC system. Simultaneously, emphasis should be given in the determination of the optimum range of mass flow rate in every case as it has large impact on the system's performance.
- Experimental investigation on the utilisation of inserting metal foams, continuous and discontinuous helical screw-tape inserts, conical rings in various angles and other insert types are needed to optimise the design of PTC systems.
- Further experimental research is needed to identify the impact of incidence angle with its bi-axial modifiers on the PTC's thermal efficiency.
- The exergetic study of a PTC is an essential means of evaluating its performance. The exergetic optimisation and assessment would provide clear indication on the exergetic losses and exergy destructions to identify the main causes of decrease in the useful exergetic output.
- Further work may be carried out on the numerical analysis of absorber tube supported at multiple points to identify its impact on temperature profile and deflection, and its effect on heat transfer into the HTF. Using various other materials as a combination to diminish the receiver's pipe weight and to augment the heat conduction can also be analysed. The decrease in heat flux value on the absorber tube surface due to deflection, and its effect on hydraulic field due to bending can also be further analysed numerically.
- Further studies on different temperature distributions, different HTF types and different residual gas conditions will be needed to investigate the characteristics of fluid dynamics/coupled heat transfer on the PTC's thermal effectiveness and overall efficiency synthetically.
- Further work on validating the thermal model results with the experimental examinations of un-irradiated PTC receivers is needed. More research is also needed to estimate the influence of the nonuniformity solar flux profile on the convective losses to the surroundings and considering the wind's angle of attack.
- The bellow, getter and the glass-to-metal seal technique that are important component of a PTC system should also be optimised in the analysis to get better understanding on the full system performance.
- It is worth noting that oxides at reduced concentration are cost-effective, therefore they could be particularly suitable for PTC experimental investigations. However, there are still various nanoparticle types, e.g., carbon nanotubes and

nanohorns, other magnetic nanoparticles and ferrofluids that could be worthy of further investigations.

- It was inferred that hybrid nanofluids are excellent alternatives for solar energy applications and are promising and intriguing thought, which is not inspected profoundly up to today, and even the studies available in the literature are very limited at this moment. Thus, a gigantic experimental and numerical work are required to execute new heat transfer fluids in sun-oriented explicit applications.
- There is an imperative need for two-phase flow research in PTC systems and it needs to be extensively explored further.
- The literature review shows that different optimum pressures are found for different inlet temperature levels in PTC analysis. Thus, the utilisation of the PTC in a Brayton cycle can be tested in the future to assess the overall system performance.
- The cross-area of the PTC's absorber pipe is suggested to be in various shapes such as elliptical for future investigations to evaluate its impacts on the thermal effectiveness.
- There are insufficient examinations for nanofluids utilisation in PTC frameworks. The coupling of passive techniques and nanofluids can have unbelievably great impacts on the PTC's thermal effectiveness. The requirement for additional examinations on the utilisation of nanofluids in permeable media is additionally suggested because of its exceptional thermal properties.
- There are different strategies for optimisation, e.g., genetic algorithm and Taguchi could also be utilised to accomplish an ideal structure of PTCs to explore the impacts of inserts, working fluids (nanofluids, permeable media) on the thermal and hydraulic effectiveness.
- This review has also highlighted that there are few papers mentioned other nanofluid properties, e.g., surface tension, photo-thermal reaction and consistency. Moreover, there are not more modelling investigations have attempted other nanofluid properties. Thus, the current review has featured the necessity to study nanofluids properties in more depth to examine their remarkable impacts on the PTC's framework overall performance.
- There is a high need for more experimental studies with simultaneous numerical analysis on PTC systems-based hybrid nanofluids in real applications.

Acknowledgements The research presented in this study did not obtain any funds from any internal or external funding agencies. Some figures included in this article have been reproduced from the given references and copyright permission have been taken from the respective publishers.

References

Abid, M., Ratlamwala, T. A. H., & Atikol, U. (2016). Performance assessment of parabolic dish and parabolic trough solar thermal power plant using nanofluids and molten salts. *International Journal of Energy Research, 40*, 550–563.

Aldali, Y., Muneer, T., & Henderson, D. (2013). Solar absorber tube analysis: Thermal simulation using CFD. *International Journal of Low-Carbon Technologies, 8*, 14–19.

Alghoul, M. A., Sulaiman, M. Y., Azmi, B. Z., & Wahab, M. A. (2005). Review of materials for solar thermal collectors. *Anti-Corrosion Methods and Materials, 52*, 199–206.

Allouhia, A., Benzakour Amine, M., Saidur, R., Kousksou, T., & Jamil, A. (2018). Energy and exergy analyses of a parabolic trough collector operated with nanofluids for medium and high temperature applications. *Energy Conversion and Management, 155*, 201–217.

Al-Madani, H. (2006). The performance of a cylindrical solar water heater. *Renewable Energy, 31*, 1751–1763.

Alsaady, M., Fu, R., Yan, Y., Liu, Z., Wu, S., & Boukhanouf, R. (2018). An experimental investigation on the effect of Ferrofluids on the efficiency of novel parabolic trough solar collector under laminar flow conditions. *Heat Transfer Engineering, 1*(1), 1–9.

Amani, E., & Nobari, M. R. H. (2011). A numerical investigation of entropy generation in the entrance region of curved pipes at constant wall temperature. *Journal of Energy, 38*(11), 1–10.

Arasu, A. V., & Sornakumar, S. T. (2006). Performance characteristics of the solar parabolic trough collector with hot water generation system. *Thermal Science, 10*(2), 167–174.

Bakos, G. C. (2006). Design and construction of a two-axis sun tracking system for parabolic trough collector (PTC) efficiency improvement. *Renewable Energy, 31*, 2411–2421.

Basbous, N., Taqi, M., & Belouaggadia, N. (2015). Numerical study of a parabolic trough collector using a nanofluid. *Asian Journal of Current Engineering and Maths, 4*(3), 40–44.

Basbous, N., Taqi, M., & Janan, M. A. (2016). *Thermal performances analysis of a parabolic trough solar collector using different nanofluids*. In IEEE, pp. 1–5.

Behar, O., Khellaf, A., & Mohammedi, K. (2013). A review on central receiver solar thermal power plants. *Renewable and Sustainable Energy Reviews, 23*, 12–39.

Behar, O., Khellaf, A., & Mohammedi, K. (2015). A novel parabolic trough solar collector model–Validation with experimental data and comparison to Engineering Equation Solver (EES). *Energy Conversion and Management, 106*, 268–281.

Bejan, A. (1996). *Entropy generation minimization: The method of thermodynamic optimization of finite-size systems and finite-time processes*. CRC Press.

Bellos, E., & Tzivanidis, C. (2017a). Parametric investigation of supercritical carbon dioxide utilization in parabolic trough collectors. *Applied Thermal Engineering, 127*, 736–747.

Bellos, E., & Tzivanidis, C. (2017b). Parametric investigation of nanofluids utilization in parabolic trough collectors. *Thermal Science and Engineering Progress, 2*, 71–79.

Bellos, E., & Tzivanidis, C. (2018a). Thermal analysis of parabolic trough collector operating with mono and hybrid nanofluids. *Sustainable Energy Technologies and Assessments, 26*, 105–115.

Bellos, E., & Tzivanidis, C. (2018b). Assessment of the thermal enhancement methods in parabolic trough collectors. *International Journal of Energy and Environmental Engineering, 9*, 59–70.

Bellos, E., & Tzivanidis, C. (2018c). Investigation of a star flow insert in a parabolic trough solar collector. *Applied Energy, 224*, 86–102.

Bellos, E., & Tzivanidis, C. (2018d). Thermal efficiency enhancement of nanofluid-based parabolic trough collectors. *Journal of Thermal Analysis and Calorimetry, 1*, 1–12.

Bellos, E., Tzivanidis, C., Antonopoulos, K. A., & Daniil, I. (2016a). The use of gas working fluids in parabolic trough collectors – An energetic and exergetic analysis. *Applied Thermal Engineering, 109*, 1–14.

Bellos, E., Tzivanidis, C., Antonopoulos, K. A., & Gkinis, G. (2016b). Thermal enhancement of solar parabolic trough collectors by using nanofluids and converging-diverging absorber tube. *Renewable Energy, 94*, 213–222.

Bellos, E., Tzivanidis, C., & Antonopoulos, K. A. (2017a). A detailed working fluid investigation for solar parabolic trough collectors. *Applied Thermal Engineering, 114*, 374–386.

Bellos, E., Tzivanidis, C., & Daniil, I. (2017b). Energetic and exergetic investigation of a parabolic trough collector with internal fins operating with carbon dioxide. *International Journal of Energy and Environmental Engineering, 8*, 109–122.

Bellos, E., Tzivanidis, C., Daniil, I., & Antonopoulos, K. A. (2017c). The impact of internal longitudinal fins in parabolic trough collectors operating with gases. *Energy Conversion and Management, 135*, 35–54.

Bellos, E., Tzivanidis, C., & Tsimpoukis, D. (2017d). Multi-criteria evaluation of parabolic trough collector with internally finned absorbers. *Applied Energy, 205*, 540–561.

Bellos, E., Tzivanidis, C., & Tsimpoukis, D. (2017e). Thermal enhancement of parabolic trough collector with internally finned absorbers. *Solar Energy, 157*, 514–531.

Bellos, E., Tzivanidis, C., & Tsimpoukis, D. (2018a). Thermal, hydraulic and exergetic evaluation of a parabolic trough collector operating with thermal oil and molten salt based nanofluids. *Energy Conversion and Management, 156*, 388–402.

Bellos, E., Tzivanidis, C., & Tsimpoukis, D. (2018b). Optimum number of internal fins in parabolic trough collectors. *Applied Thermal Engineering, 137*, 669–677.

Bellos, E., Tzivanidis, C., & Tsimpoukis, D. (2018c). Enhancing the performance of parabolic trough collectors using nanofluids and turbulators. *Renewable and Sustainable Energy Reviews, 91*, 358–375.

Benabderrahmane, A., Aminallah, M., Laouedj, S., Benazza, A., & Solano, J. P. (2016). Heat transfer enhancement in a parabolic trough solar receiver using longitudinal fins and nanofluids. *Journal of Thermal Science, 25*(5), 410–417.

Bitam, E., Demagh, Y., Hachicha, A. A., Benmoussa, H., & Kabar, Y. (2018). Numerical investigation of a novel sinusoidal tube receiver for parabolic trough technology. *Applied Energy, 218*, 494–510.

Blanco, J., Malato, S., Fernández-Ibañez, P., Alarcón, D., Gernjak, W., & Maldonado, M. (2009). Review of feasible solar energy applications to water processes. *Renew Sust Energy Reviews, 13*(6–7): 1437–1445.

Bortolato, M., Dugaria, S., & Del Col, D. (2016). Experimental study of a parabolic trough solar collector with flat bar-and-plate absorber during direct steam generation. *Energy, 116*, 1039–1050.

Bortolato, M., Dugaria, S., Agresti, F., Barison, S., Fedele, L., Sani, E., & Del Col, D. (2017). Investigation of a single wall carbon nanohorn-based nanofluid in a fullscale direct absorption parabolic trough solar collector. *Energy Conversion and Management, 150*, 693–703.

Brooks, M. J., Mills, I., & Harms, T. M. (2006). Performance of a parabolic trough solar collector. *Journal of Energy in Southern Africa, 17*(3), 71–80.

Cameron, C. P., & Dudley, V. E. (1987). *Acurex solar corporation modular industrial solar retrofit qualification test results* (Technical report no. SAND85-2316). SANDIA.

Cameron, C. P., Dudley, D. V. E., & Lewandowski, A. A. (1986). *Foster wheeler solar development corporation modular industrial solar retrofit qualification test results* (Technical report no. SAND85-2319). SANDIA.

Carlton, R. J. (1981). *Tracking solar energy collector assembly. Acurex corporation* (Patent no. 4297572). The Patent Office of London.

Chafie, M., Ben Aissa, M. F., & Guizani, A. (2018). Energetic end exergetic performance of a parabolic trough collector receiver: An experimental study. *Journal of Cleaner Production, 171*, 285–296.

Chang, C., Li, X., & Zhang, Q. Q. (2014). Experimental and numerical study of the heat transfer characteristics in solar thermal absorber tubes with circumferentially non-uniform heat flux. *Energy Procedia, 49*, 305–313.

Chang, C., Xu, C., Wu, Z. Y., Li, X., Zhang, Q. Q., & Wang, Z. F. (2015). Heat transfer enhancement and performance of solar thermal absorber tubes with circumferentially non-uniform heat flux. *Energy Procedia, 69*, 320–327. https://doi.org/10.1016/j.egypro.2015.03.036

Chang, C., Peng, X., Nie, B., Leng, G., Li, C., Hao, Y., & She, X. (2017). Heat transfer enhancement of a molten salt parabolic trough solar receiver with concentric and eccentric pipe inserts. *Energy Procedia, 142*, 624–629.

Chang, C., Sciacovelli, A., Wu, Z., Li, X., Li, Y., Zhao, M., Deng, J., Wang, Z., & Ding, Y. (2018). Enhanced heat transfer in a parabolic trough solar receiver by inserting rods and using molten salt as heat transfer fluid. *Applied Energy, 220*, 337–350.

Channiwala, S. A., & Ekbote, A. (2015). *A generalized model to estimate field size for solar-only parabolic trough plants*. In Proceedings of the 3rd Southern African Solar Energy Conference (SASEC).

Chaudhari, K. S., Walke, P. V., Wankhede, U. S., & Shelke, R. S. (2015). An experimental investigation of a nanofluid (Al_2O_3+H_2O) based parabolic trough solar collectors. *British Journal of Applied Science & Technology, 9*(6), 551–557.

Chean, Y., Too, S., & Benito, R. (2013). Enhancing heat transfer in air tubular absorbers for concentrated solar thermal applications. *Applied Thermal Engineering, 50*, 1076–1083.

Cheng, Z. D., He, Y. L., Xiao, J., Tao, Y. B., & Xu, R. J. (2010). Three-dimensional numerical study of heat transfer characteristics in the receiver tube of parabolic trough solar collector. *International Communications in Heat and Mass Transfer, 37*, 782–787.

Cheng, Z. D., He, Y. L., & Cui, F. Q. (2012a). Numerical study of heat transfer enhancement by unilateral longitudinal vortex generators inside parabolic trough solar receivers. *International Journal of Heat and Mass Transfer, 55*, 5631–5641.

Cheng, Z. D., He, Y. L., Cui, F. Q., Xu, R. J., & Tao, Y. B. (2012b). Numerical simulation of a parabolic trough solar collector with nonuniform solar flux conditions by coupling FVM and MCRT method. *Solar Energy, 86*, 1770–1784.

Cheng, Z. D., He, Y. L., Cui, F. Q., Du, B. C., Zheng, Z. J., & Xu, Y. (2014). Comparative and sensitive analysis for parabolic trough solar collectors with a detailed Monte Carlo ray-tracing optical model. *Applied Energy, 115*, 559–572.

Cheng, Z. D., He, Y. L., & Qiu, Y. (2015). A detailed nonuniform thermal model of a parabolic trough solar receiver with two halves and two inactive ends. *Renewable Energy, 74*, 139–147.

Choi, S. U. S. (1995). Enhancing thermal conductivity of fluids with nanoparticles, In D. A. Siginer & H. P. Wang (Eds.). *Developments and applications of non-Newtonian flows, in: FED, vol. 231/MD, vol. 66* (pp. 99–103). ASME.

Coccia, G., Di Nicola, G., Colla, L., Fedele, L., & Scattolini, M. (2016). Adoption of nanofluids in low-enthalpy parabolic trough solar collectors: Numerical simulation of the yearly yield. *Energy Conversion and Management, 118*, 306–319.

Conrado, L. S., Rodriguez-Pulido, A., Calderón, G., & G. (2017). Thermal performance of parabolic trough solar collectors. *Renewable and Sustainable Energy Reviews, 67*, 1345–1359.

CSP Technologies. (2018). Retrieved from https://hub.globalccsinstitute.com/publications/concentrating-solar-power-india/21-csp-technologies. Accessed 10 July 2018.

Das, S. K., Choi, S. U. S., & Patel, H. P. (2006). Heat transfer in nanofluids: A review. *Heat Transfer Engineering, 27*, 3–19.

De los Rios, M. S. B., Rivera-Solorio, C. I., & García-Cuellar, A. J. (2018). Thermal performance of a parabolic trough linear collector using Al_2O_3/H_2O nanofluids. *Renewable Energy, 122*, 665–673.

de Risi, A., Milanese, M., & Laforgia, D. (2013). Modelling and optimization of transparent parabolic trough collector based on gas-phase nanofluids. *Renewable Energy, 58*, 134–139.

Dean, J. A. (1992). *Lange's handbook of chemistry* (14th ed.). McGraw-Hill.

Dewan, A., Mahanta, P., Sumithra Raju, K., & Suresh Kumar, P. (2004). Review of passive heat transfer augmentation techniques. *Proceedings of the I MECH E Part A Journal of Power and Energy, 218*, 509–527.

Diwan, K., & Soni, M. S. (2015). Heat transfer enhancement in absorber tube of parabolic trough concentrators using wire-coils inserts. *Universal Journal of Mechanical Engineering, 3*(3), 107–112.

Dudley, V. E., & Workhoven, R. M. (1981). *Performance testing of the Acurex solar collector model 3001-03* (Technical report no. SAND80-0872). SANDIA.

Duffie, J. A., & Beckman, W. A. (1991). *Solar engineering of thermal processes* (2nd ed.). Wiley.

Duncan, A. B., & Peterson, G. P. (1994). Review of micro-scale heat transfer. *Applied Mechanics Reviews, 47*, 397–428.

Eastop, T. D., & McConkey, A. (1977). *Applied thermodynamics for engineering technologists.* Longman.

Edenburn, M. W. (1976). Performance analysis of a cylindrical parabolic focusing collector and comparison with experimental results. *Solar Energy, 18*, 437–444.

Ericsson, J. (1884). The sun motor and the Sun's temperature. *Nature, 29*, 217–223.

Essentials, R. E. (2009). *Concentrating solar thermal power.* International Energy Agency (IEA).

Fan, M., Liang, H., You, S., Zhang, H., Zheng, W., & Xia, J. (2018). Heat transfer analysis of a new volumetric based receiver for parabolic trough solar collector. *Energy, 142*, 920–931.

Fernandez-Garcia, A., Zarza, E., Valenzuela, L., & Perez, M. (2010). Parabolic-trough solar collectors and their applications. *Renewable and Sustainable Energy Reviews, 14*, 1695–1721.

Fernández-García, A., Rojas, E., Pérez, M., Silva, R., Hernández-Escobedo, O., & Manzano-Agugliaro, F. (2015). A parabolic-trough collector for cleaner industrial process heat. *Journal of Cleaner Production, 89*, 272–285.

Ferraroa, V., Settino, J., Cucumo, M. A., & Kaliakatsos, D. (2016). Parabolic trough system operating with nanofluids: Comparison with the conventional working fluids and influence on the system performance. *Energy Procedia, 101*, 782–789. https://doi.org/10.1016/j.egypro.2016.11.099

Fuqiang, W., Qingzhi, L., Huaizhi, H., & Jianyu, T. (2016a). Parabolic trough receiver with corrugated tube for improving heat transfer and thermal deformation characteristics. *Applied Energy, 164*, 411–424.

Fuqiang, W., Xiangtao, G., Jianyu, T., Huaizhi, H., & Bingxi, L. (2016b). Heat transfer performance enhancement and thermal strain restrain of tube receiver for parabolic trough solar collector by using asymmetric outward convex corrugated tube. *Energy, 114*, 275–292.

Gaul, H., & Rabl, A. (1980). Incidence-angle modifier and average optical efficiency of parabolic trough collectors. *Journal of Solar Energy Engineering, 102*, 16–21.

Geyer, P. E., Fletcher, D. F., & Haynes, B. S. (2007). Laminar flow and heat transfer in a periodic trapezoidal channel with semi-circular cross section. *International Journal of Heat and Mass Transfer, 50*, 3471–3480.

Ghadirijafarbeigloo, S., Zamzamian, A. H., & Yaghoubi, M. (2014). 3-D numerical simulation of heat transfer and turbulent flow in a receiver tube of solar parabolic trough concentrator with louvered twisted-tape inserts. *Energy Procedia, 49*, 373–380.

Ghasemi, S. E., & Mehdizadeh Ahangar, G. H. R. (2014). Numerical analysis of performance of solar parabolic trough collector with Cu-water nanofluid. *International Journal of Nano Dimension, 5*(3), 233–240.

Ghasemi, S. E., & Ranjbar, A. A. (2016). Thermal performance analysis of solar parabolic trough collector using nanofluid as working fluid: A CFD modelling study. *Journal of Molecular Liquids, 222*, 159–166.

Ghasemi, S. E., & Ranjbar, A. A. (2017a). Effect of using nanofluids on efficiency of parabolic trough collectors in solar thermal electric power plants. *International Journal of Hydrogen Energy, 42*, 21626–21634.

Ghasemi, S. E., & Ranjbar, A. A. (2017b). Numerical thermal study on effect of porous rings on performance of solar parabolic trough collector. *Applied Thermal Engineering, 118*, 807–816.

Ghasemi, S. E., Ranjbar, A. A., & Ramiar, A. (2013). Three-dimensional numerical analysis of heat transfer characteristics of solar parabolic collector with two segmental rings. *Journal of Mathematics and Computer Science, 7*, 89–100.

Godson, L., Raja, B., Mohan, L. D., & Wongwises, S. (2010). Experimental investigation on the thermal conductivity and viscosity of silver-deionized water nanofluid. *Experimental Heat Transfer, 23*, 317–332.

Gouthamraj, K., Rani, K., & Satyanarayana, G. (2013). Design and analysis of rooftop linear fresnel reflector solar concentrator. *International Journal of Engineering and Innovative Technology (IJEIT), 2*, 66–69.

Grald, E. W., & Kuehn, T. H. (1989). Performance analysis of a parabolic trough solar collector with a porous absorber receiver. *Solar Energy, 42*(4), 281–292.

Grasse, W., Hertlein, H., Winter, C. J., & Braun, G. (1991). Thermal solar power plants experience. In *Solar power plants*. Springer.

Günther, M., Joemann, M., & Csambor, S. (2011). Chapter 5: Parabolic trough technology. In *Advanced CSP teaching materials*. Deutsches Zentrums für Luft- und Raumfahrt.

Guo, J., Huai, X., & Liu, Z. (2016). Performance investigation of parabolic trough solar receiver. *Applied Thermal Engineering, 95*, 357–364.

Guven, H., & Bannerot, R. (1986). Derivation of universal error parameters for comprehensive optical analysis of parabolic troughs. *Journal of Solar Energy Engineering, 108*, 275–281.

Hachicha, A. A., Rodríguez, I., Capdevila, R., & Oliva, A. (2013). Heat transfer analysis and numerical simulation of a parabolic trough solar collector. *Applied Energy, 111*, 581–592.

He, Y. L., Xiao, J., Cheng, Z. D., & Tao, Y. B. (2011). A MCRT and FVM coupled simulation method for energy conversion process in parabolic trough solar collector. *Renewable Energy, 36*, 976–985.

Hellstrom, B., Adsten, M., Nostell, P., Karlsson, B., & Wackelgard, E. (2003). The impact of optical and thermal properties on the performance of flat plate solar collectors. *Renewable Energy, 28*, 331–344.

Herwig, H., & Kock, F. (2007). Direct and indirect methods of calculating entropy generation rates in turbulent convective heat transfer problems. *Journal of Heat and Mass Transfer, 43*, 207–215.

Hojjat, M., Etemad, S. G., Bagheri, R., & Thibault, J. (2011). Turbulent forced convection heat transfer of non-Newtonian nanofluids. *Experimental Thermal and Fluid Science, 35*, 1351–1356.

Hsieh, S. S., & Huang, I. W. (2000). Heat transfer and pressure drop of laminar flow in horizontal tubes with/without longitudinal inserts. *Journal of Heat Transfer, 122*, 465–475.

Huang, Z., Yu, G. L., Li, Z. Y., & Tao, W. Q. (2015). Numerical study on heat transfer enhancement in a receiver tube of parabolic trough solar collector with dimples, protrusions and helical fins. *Energy Procedia, 69*, 1306–1316.

Huang, Z., Li, Z. Y., Yu, G. L., & Tao, W. Q. (2017). Numerical investigations on fully-developed mixed turbulent convection in dimpled parabolic trough receiver tubes. *Applied Thermal Engineering, 114*, 1287–1299.

Hwang, Y., Lee, J. K., Lee, C. H., Jung, Y. M., Cheong, S. I., Lee, C. G., Ku, B. C., & Jang, S. P. (2007). Stability and thermal conductivity characteristics of nanofluids. *Thermochimica Acta, 455*, 70–74.

International Energy Agency. (2009). *Technology roadmap: Concentrating solar power.* Paris (FR): IEA Publications.

International Renewable Energy Agency. (2012). *Renewable energy technologies: Cost analysis series (concentrating solar power).* IRENA Publications [Cited 19 Aug 2020]. Available from: https://www.irena.org/-/media/Files/IRENA/Agency/Publication/2012/RE_Technologies_Cost_Analysis-CSP.pdf

International Renewable Energy Agency. (2015). *Renewable power generation costs in 2014.* IRENA Publications [Cited 2020 Aug 19]. Available from: https://www.irena.org/-/media/Files/IRENA/Agency/Publication/2015/IRENA_RE_Power_Costs_2014_report.pdf

Islam, M., Karim, A., Saha, A. C., Miller, S., Prasad, K. D., & Yarlagadda, V. (2012). *Three dimensional simulation of a parabolic trough concentrator thermal collector.* In Proceedings of the 50th annual conference. Australian Solar Energy Society (Australian Solar Council), Swinburne University of Technology, Melbourne, Australia.

Jamal-Abad, M. T., Saedodin, S., & Aminy, M. (2017). Experimental investigation on a solar parabolic trough collector for absorber tube filled with porous media. *Renewable Energy, 107*, 156–163.

Jaramillo, O. A., Borunda, M., Velazquez-Lucho, K. M., & Robles, M. (2016). Parabolic trough solar collector for low enthalpy processes: An analysis of the efficiency enhancement by using twisted tape inserts. *Renewable Energy, 93*, 125–141.

Jebasingh, V. K., & Joselin Herbert, G. M. (2016). A review of solar parabolic trough collector. *Renewable and Sustainable Energy Reviews, 54*, 1085–1091.

Kainth, M., & Sharma, V. K. (2014). Latest evolutions in flat plate solar collectors technology. *International Journal of Mechanical Engineering, 1*, 7–11.

Kalogirou, S. A. (2004). Solar thermal collectors and applications. *Progress in Energy and Combustion Science, 30*, 231–295.

Kalogirou, S. A. (2012). A detailed thermal model of a parabolic trough collector receiver. *Energy, 48*(1), 298–306.

Kaloudis, E., Papanicolaou, E., & Belessiotis, V. (2016). Numerical simulations of a parabolic trough solar collector with nanofluid using a two-phase model. *Renewable Energy, 97*, 218–229.

Kalt, A., Loosme, M., & Dehne, H. (1981). *Distributed collector system plant construction* (Technical report no. IEA-SSPS-SR-1). IEA.

Kasaeian, A. B., Sokhansefat, T., Abbaspour, M. J., & Sokhansefat, M. (2012). Numerical study of heat transfer enhancement by using Al_2O_3/synthetic oil Nanofluid in a parabolic trough collector tube. *World Academy of Science, Engineering and Technology, 69*, 1154–1159.

Kasaeian, A., Daviran, S., Azarian, R. D., & Rashidi, A. (2015). Performance evaluation and nanofluid using capability study of a solar parabolic trough collector. *Energy Conversion and Management, 89*, 368–375.

Kasaeian, A., Daneshazarian, R., & Pourfayaz, F. (2017). Comparative study of different nanofluids applied in a trough collector with glass-glass absorber tube. *Journal of Molecular Liquids, 234*, 315–323.

Kesselring, P., & Selvage, C. S. (1986). *The IEA/SSPS solar thermal power plants* (vol. 2: Distributed collector system (DCS)) (1st ed.). Springer.

Khakrah, H., Shamloo, A., & Hannani, S. K. (2017). Determination of parabolic trough solar collector efficiency using Nanofluid: A comprehensive numerical study. *Journal of Solar Energy Engineering, 139*, 051006-051001–051006-051011.

Khan, M. S., Yan, M., Ali, H. M., Amber, K. P., Bashir, M. A., Akbar, et al. (2020). Comparative performance assessment of different absorber tube geometries for parabolic trough solar collector using nanofluid. *Journal of Thermal Analysis and Calorimetry, 142*, 2227–2241.

Khullar, V., Tyagi, H., Phelan, P. E., Otanicar, T. P., Singh, H., & Taylor, R. A. (2012). Solar energy harvesting using nanofluids-based concentrating solar collector. *Journal of Nanotechnology in Engineering and Medicine, 3,* 031003-031001–031003-031009.

Kim, S. H., Choi, S. R., & Kim, D. (2007). Thermal conductivity of metal-oxide nanofluids: Particle size dependence and effect of laser irradiation. *Journal of Heat Transfer, 129,* 298–307.

Konttinen, P., Lund, P. D., & Kilpi, R. J. (2003). Mechanically manufactured selective solar absorber surfaces. *Solar Energy Materials & Solar Cells, 79,* 273–283.

Kreider, J. F., & Kreith, F. (1981). *Solar energy handbook.* McGraw Hill.

Kumar, D., & Kumar, S. (2018). Thermal performance of the solar parabolic trough collector at different flow rates: An experimental study. *International Journal of Ambient Energy, 39*(1), 93–102.

Kumar, K. R., & Reddy, K. S. (2009). Thermal analysis of solar parabolic trough with porous disc receiver. *Applied Energy, 86,* 1804–1812.

Kumar, K. R., & Reddy, K. S. (2012). Effect of porous disc receiver configurations on performance of solar parabolic trough concentrator. *Heat and Mass Transfer, 48,* 555–571.

Kumar, M. K., Yuvaraj, G., Balaji, D., Pravinraj, R., & Shanmugasundaram, P. (2018). Influence of nano-fluid and receiver modification in solar parabolic trough collector performance. *IOP Conference Series: Materials Science and Engineering, 310,* 012140.

Kumaresan, G., Sridhar, R., & Velraj, R. (2012). Performance studies of a solar parabolic trough collector with a thermal energy storage system. *Energy, 47*(1), 395–402.

Kutscher, C. F., DavenportL, R. L., Dougherty, D. A., Gee, R. C., Masterson, P. M., & May, E. K. (1982). *Design approaches for solar industrial process heat systems* (Technical report no. SERI/TR-253-1358). SERI.

Lee, D. W., & Sharma, A. (2007). Thermal performances of the active and passive water heating systems based on annual operation. *Solar Energy, 81,* 207–215.

Li, Z. Y., Huang, Z., & Tao, W. Q. (2015). Three-dimensional numerical study on turbulent mixed convection in parabolic trough solar receiver tube. *Energy Procedia, 75,* 462–466.

Li, Z. Y., Huang, Z., & Tao, W. Q. (2016). Three-dimensional numerical study on fully-developed mixed laminar convection in parabolic trough solar receiver tube. *Energy, 113,* 1288–1303.

Liu, Q., Wang, Y., Gao, Z., Sui, J., Jin, H., & Li, H. (2010). Experimental investigation on a parabolic trough solar collector for thermal power generation. *Science in China Series E: Technological Sciences, 53*(1), 52–56.

Lobón, D. H., & Valenzuela, L. (2013). Impact of pressure losses in small-sized parabolic-trough collectors for direct steam generation. *Energy, 61,* 502–512.

Lu, J., Ding, J., Yu, T., & Shen, X. (2015). Enhanced heat transfer performances of molten salt receiver with spirally grooved pipe. *Applied Thermal Engineering, 88,* 491–498.

Lu, J., Yuan, Q., Ding, J., Wang, W., & Liang, J. (2016). Experimental studies on nonuniform heat transfer and deformation performances for trough solar receiver. *Applied Thermal Engineering, 109,* 497–506.

Lüpfert E, Geyer M, Schiel W, Esteban A, Osuna R, Zarza E, et al. (2001). *EuroTrough design issues and prototype testing at PSA.* In Proceedings of solar forum. ASME.

Lüpfert, E., Zarza, E., Geyer, M., Nava, P., Langenkamp, J., Schiel, W., et al. (2003). *EuroTrough collector qualification complete-performance test results from PSA.* ISES Solar World Congress.

Maier, W., & Remshardt, A. (1907). *Vorrichtung zur unmittelbaren Verwendung der Sonnenwärme zur Dampferzeugung* (Patent Nr. 231294). Kaiserliches Patentamt.

Malato, S., Blanco, J., Maldonado, M., Fernández-Ibañez, P., Alarcón, D., Collares, M., et al. (2004). Engineering of solar photocatalysis collectors. *Solar Energy, 77*(5): 513–524.

Manikandan, K. S., Kumaresan, G., Velraj, R., & Iniyan, S. (2012). Parametric study of solar parabolic trough collector system. *Asian Journal of Applied Sciences, 5,* 384–393.

Marefati, M., Mehrpooya, M., & Shafii, M. B. (2018). Optical and thermal analysis of a parabolic trough solar collector for production of thermal energy in different climates in Iran with comparison between the conventional nanofluids. *Journal of Cleaner Production, 175,* 294–313.

Menbari, A., Alemrajabi, A. A., & Rezaei, A. (2017). Experimental investigation of thermal performance for direct absorption solar parabolic trough collector (DASPTC) based on binary nanofluids. *Experimental Thermal and Fluid Science, 80*, 218–227.

Minardi, J. E., & Chuang, H. N. (1975). Performance of a "black" liquid flat-plate solar collector. *Solar Energy, 17*, 179–183.

Minea, A. A., & El-Maghlany, W. M. (2018). Influence of hybrid nanofluids on the performance of parabolic trough collectors in solar thermal systems: Recent findings and numerical comparison. *Renewable Energy, 120*, 350–364.

Mokheimer, E. M., Dabwan, Y. N., Habib, M. A., Said, S. A., & Al-Sulaiman, F. A. (2014). Techno-economic performance analysis of parabolic trough collector in Dhahran, Saudi Arabia. *Energy Conversation and Management, 86*, 622–633.

Muñoz, J., & Abánades, A. (2011). Analysis of internal helically finned tubes for parabolic trough design by CFD tools. *Applied Energy, 88*, 4139–4149.

Mwesigye, A., & Huan, Z. (2015). Thermal and thermodynamics performance of a parabolic trough receiver with Syltherm 800-Al_2O_3 nanofluid as the heat transfer fluid. *Energy Procedia, 75*, 394–402.

Mwesigye, A., & Meyer, J. P. (2017). Optimal thermal and thermodynamic performance of a solar parabolic trough receiver with different nanofluids and at different concentration ratios. *Applied Energy, 193*, 393–413.

Mwesigye, A., Bello-Ochende, T., & Meyer, J. P. (2013). Numerical investigation of entropy generation in a parabolic trough receiver at different concentration ratios. *Energy, 53*, 114–127.

Mwesigye, A., Bello-Ochende, T., & Meyer, J. P. (2014a). Heat transfer and thermodynamic performance of a parabolic trough receiver with centrally placed perforated plate inserts. *Applied Energy, 136*, 989–1003.

Mwesigye, A., Bello-Ochende, T., & Meyer, J. P. (2014b). Minimum entropy generation due to heat transfer and fluid friction in a parabolic trough receiver with non-uniform heat flux at different rim angles and concentration ratios. *Energy, 73*, 606–617.

Mwesigye, A., Bello-Ochende, T., & Meyer, J. P. (2015). Multi-objective and thermodynamic optimisation of a parabolic trough receiver with perforated plate inserts. *Applied Thermal Engineering, 77*, 42–56.

Mwesigye, A., Huan, Z., Meyer, J. P. (2015a). *Thermal performance of a receiver tube for a high concentration ratio parabolic trough system and potential for improved performance with syltherm800-CuO nanofluid*. In Proceedings of the ASME 2015 International Mechanical Engineering Congress & Exposition IMECE2015, Houston, Texas, USA.

Mwesigye, A., Huan, Z., & Meyer, J. P. (2015b). Thermodynamic optimisation of the performance of a parabolic trough receiver using synthetic oil–Al_2O_3 nanofluid. *Applied Energy, 156*, 398–412.

Mwesigye, A., Bello-Ochende, T., & Meyer, J. P. (2016a). Heat transfer and entropy generation in a parabolic trough receiver with wall-detached twisted tape inserts. *International Journal of Thermal Sciences, 99*, 238–257.

Mwesigye, A., Huan, Z., & Meyer, J. P. (2016b). Thermal performance and entropy generation analysis of a high concentration ratio parabolic trough solar collector with Cu-Therminol VP-1 nanofluid. *Energy Conversion and Management, 120*, 449–465.

Mwesigye, A., Yılmaz, I. H., & Meyer, J. P. (2018). Numerical analysis of the thermal and thermodynamic performance of a parabolic trough solar collector using SWCNTs-Therminol VP-1 nanofluid. *Renewable Energy, 119*, 844–862.

Nems, M., & Kasperski, J. (2016). Experimental investigation of concentrated solar air-heater with internal multiple-fin array. *Renewable Energy, 97*, 722–730.

Nixon, J. D., Dey, P. K., & Davies, P. A. (2010). Which is the best solar thermal collection technology for electricity generation in North-West India? Evaluation of options using the analytical hierarchy process. *Energy, 35*, 5230–5240.

Okonkwo, E. C., Abid, M., & Ratlamwala, T. A. H. (2018). Effects of synthetic oil nanofluids and absorber geometries on the exergetic performance of the parabolic trough collector. *International Journal of Energy Research, 42*(5), 1–16.

Padilla, R. V., Fontalvo, A., Demirkaya, G., Martinez, A., & Quiroga, A. G. (2014). Exergy analysis of parabolic trough solar receiver. *Applied Thermal Engineering, 67*(1), 579–586.

Pavlovic, S., Bellos, E., Stefanovic, V., & Tzivanidis, C. (2017a). Optimum geometry of parabolic trough collectors with optical and thermal criteria. *International Review of Applied Sciences and Engineering, 8*(1), 45–50.

Pavlovic, S., Daabo, A. M., Bellos, E., Stefanovi, V., Mahmoud, S., & Al-Dadah, R. K. (2017b). Experimental and numerical investigation on the optical and thermal performance of solar parabolic dish and corrugated spiral cavity receiver. *Journal of Cleaner Production, 150*, 75–92.

Perry, R. H., & Green, D. W. (1997). *Perry's chemical engineers' handbook* (7th ed.). McGraw-Hill.

Price, H., Lupfert, E., Kearney, D., Zarza, E., Cohen, G., & Gee, R. (2002). Advances in parabolic trough solar power technology. *Journal of Solar Energy Engineering, 124*, 109–125.

Pytlinski, J. T. (1978). Solar energy installations for pumping irrigation water. *Solar Energy, 21*, 255–258.

Rabl, A. (1976). Optical and thermal properties of compound parabolic collectors. *Solar Energy, 18*, 497–511.

Ranga Babu, J. A., Kumar, K. K., & Rao, S. S. (2017). State-of-art review on hybrid nanofluids. *Renewable and Sustainable Energy Reviews, 77*, 551–565.

Rawani, A., Sharma, S. P., & Singh, K. D. P. (2017). Enhancement in performance of parabolic trough collector with serrated twisted-tape inserts. *International Journal of Thermodynamics, 20*(2), 111–119.

Rawani, A., Sharma, S. P., & Singh, K. D. P. (2018a). Performance enhancement of parabolic trough collector with Oblique Delta-winglet twisted-tape inserts. *International Energy Journal, 18*, 39–52.

Rawani, A., Sharma, S. P., & Singh, K. D. P. (2018b). Comparative performance analysis of different twisted tape inserts in the absorber tube of parabolic trough collector. *International Journal of Mechanical and Production Engineering Research and Development, 8*(1), 643–656.

Reddy, K. S., & Satyanarayana, G. V. (2008). Numerical study of porous finned receiver for solar parabolic trough concentrator. *Engineering Applications of Computational Fluid Mechanics, 2*(2), 72–184.

Reddy, K. S., Kumar, K. R., & Satyanarayana, G. V. (2008). Numerical investigation of energy-efficient receiver for solar parabolic trough concentrator. *Heat Transfer Engineering, 29*(11), 961–972.

Reddy, K. S., Kumar, K. R., & Ajay, C. S. (2015). Experimental investigation of porous disc enhanced receiver for solar parabolic trough collector. *Renewable Energy, 77*, 308–319.

Rehan, M. A., Ali, M., Sheikh, N. A., Khalil, M. S., Chaudhary, G. Q., Ur Rashid, T., & Shehryar, M. (2018). Experimental performance analysis of low concentration ratio solar parabolic trough collectors with nanofluids in winter conditions. *Renewable Energy, 118*, 742–751.

Rifflemann, K., Richert, T., Nava, P., & Schweitzer, A. (2014). Ultimate trough: A significant step towards cost- competitive CSP. *Energy Procedia, 49*, 1831–1839.

Romero, M., Martınez, D., & Zarza, E. (2004). *Terrestrial solar thermal power plants: On the verge of commercialization.* In 4th International conference on solar power from space.

Rotemi. (2009). http://www.rotemi.co.il/text/upload/pdf2/cleantech/SolarDesal.pdf

Sadaghiyani, O. K., Pesteei, S. M., & Mirzaee, I. (2014). Numerical study on heat transfer enhancement and friction factor of LS-2 parabolic solar collector. *Journal of Thermal Science and Engineering Applications, 6*(012001), 1–10.

Sadaghiyani, O. K., Boubakran, M. S., & Hassanzadeh, A. (2018). Energy and exergy analysis of parabolic trough collectors. *International Journal of Heat and Technology, 36*(1), 147–158.

Saidur, R., Leong, K. Y., & Mohammed, H. A. (2011). A review on applications and challenges of nanofluids. *Renew Sust Energy Reviews, 15*(3), 1646–1668.

Sarbu, I., & Sebarchievici, C. (2013). Review of solar refrigeration and cooling systems. *Energy and Buildings, 67*, 286–297.

Shaner, W. W., & Duff, W. S. (1979). Solar thermal electric power systems: Comparison of line-focus collectors. *Solar Energy, 22*, 13–49.

Shou, C. H., Luo, Z. Y., Wang, T., Cai, J. C., Zhao, J. F., Ni, M. J., & Cen, K. F. (2009). Research on the application of nano-fluids into the solar photoelectric utilization. *Shanghai Electric Power, 16*, 8–12.

Shuman, F., & Boys, C. V. (1917, September 25). *Sun boiler* (Patent no. 1240890). United States Patent Office.

Sidik, N. A. C., Adamu, I. M., Jamil, M. M., Kefayati, G. H. R., Mamat, R., & Najafi, G. (2016). Recent progress on hybrid nanofluids in heat transfer applications: A comprehensive review. *International Communications in Heat and Mass Transfer, 75*, 68–79.

Sokhansefat, T., Kasaeian, A. B., & Kowsary, F. (2014). Heat transfer enhancement in parabolic trough collector tube using Al2O3/synthetic oil nanofluid. *Renewable and Sustainable Energy Reviews, 33*, 636–644.

Solel. (2009). http://www.solel.com.

Song, X., Dong, G., Gao, F., Diao, X., Zheng, L., & Zhou, F. (2014). A numerical study of parabolic trough receiver with nonuniform heat flux and helical screw-tape inserts. *Energy, 77*, 771–782.

Spencer, L. C. (1989). A comprehensive review of small solar-powered heat engines: Part I. A history of solar-powered devices up to 1950. *Solar Energy, 43*(4), 197–214.

Subramani, J., Nagarajan, P. K., Mahian, O., & Sathyamurthy, R. (2018). Efficiency and heat transfer improvements in a parabolic trough solar collector using TiO_2 nanofluids under turbulent flow regime. *Renewable Energy, 119*, 19–31.

Sukhatme, K., & Sukhatme, S. P. (1996). *Solar energy-principles of thermal collection and storage*. Tata McGraw-Hill Education.

Sunil, K., Kundan, L., & Sumeet, S. (2014). Performance evaluation of a nanofluid based parabolic solar collector – An experimental study. *International Journal of Mechanical And Production Engineering, 2*(10), 61–67.

Sureshkumar, R., Mohideen, S. T., & Nethaji, N. (2013). Heat transfer characteristics of nanofluids in heat pipes: A review. *Renewable and Sustainable Energy Reviews, 20*, 397–410.

Syed Jafar, K., & Sivaraman, B. (2014). Thermal performance of solar parabolic trough collector using nanofluids and the absorber with nail twisted tapes inserts. *International Energy, 14*, 189–198.

Tagle-Salazar, P. D., Nigam, K. D. P., & Rivera-Solorio, C. I. (2018). Heat transfer model for thermal performance analysis of parabolic trough solar collectors using nanofluids. *Renewable Energy, 125*, 334–343.

Tagle-Salazar, P. D., Nigam, K. D. P., & Rivera-Solorio, C. I. (2020). Parabolic trough solar collectors: A general overview of technology, industrial applications, energy market, modeling, and standards. *Green Processing and Synthesis, 9*, 595–649.

Tao, Y. B., & He, Y. L. (2010). Numerical study on coupled fluid flow and heat transfer process in parabolic trough solar collector tube. *Solar Energy, 84*, 1863–1872.

Tchobanoglous, G., Burton, F., & Stensel, H. (2003). *Wastewater engineering: treatment and reuse* (4th ed.). Boston: McGraw-Hill, p. 1197.

Thirugnanasambandam, M., Iniyan, S., & Goic, R. (2010). A review of solar thermal technologies. *Renewable and Sustainable Energy Reviews, 14*, 312–322.

Tian, Y., & Zhao, C. Y. (2013). A review of solar collectors and thermal energy storage in solar thermal applications. *Applied Energy, 104*, 538–553.

Too, Y. C. S., & Benito, R. (2013). Enhancing heat transfer in air tubular absorbers for concentrated solar thermal applications. *Applied Thermal Engineering, 50*(1), 1076–1083.

Tripathy, A. K., Ray, S., Sahoo, S. S., & Chakrabarty, S. (2018). Structural analysis of absorber tube used in parabolic trough solar collector and effect of materials on its bending: A computational study. *Solar Energy, 163*, 471–485.

Tzivanidis, C., Bellos, E., Korres, D., Antonopoulos, K. A., & Mitsopoulos, G. (2015). Thermal and optical efficiency investigation of a parabolic trough collector. *Case Studies in Thermal Engineering, 6*, 226–237.

Valenzuela, L., López-Martín, R., & Zarza, E. (2014). Optical and thermal performance of large-size parabolic-trough solar collectors from outdoor experiments: A test method and a case study. *Energy, 70*, 456–464.

Vasiliev, L. L. (2005). Heat pipes in modern heat exchangers. *Applied Thermal Engineering, 25*, 1–19.

Waghole, D. R., Warkhedkar, R. M., Kulkarni, V. S., & Shrivastva, R. K. (2014). Experimental investigations on heat transfer and friction factor of silver nanofliud in absorber/receiver of parabolic trough collector with twisted tape inserts. *Energy Procedia, 45*, 558–567.

Wang, P., Liu, D. Y., & Xu, C. (2013). Numerical study of heat transfer enhancement in the receiver tube of direct steam generation with parabolic trough by inserting metal foams. *Applied Energy, 102*, 449–460.

Wang, Y., Liu, Q., Lei, J., & Jin, H. (2014). A three-dimensional simulation of a parabolic trough solar collector system using molten salt as heat transfer fluid. *Applied Thermal Engineering, 70*, 462–476.

Wang, Y., Liu, Q., Lei, J., & Jin, H. (2015). Performance analysis of a parabolic trough solar collector with non-uniform solar flux conditions. *International Journal of Heat and Mass Transfer, 82*, 236–249.

Wang, Y., Xu, J., Liu, Q., Chen, Y., & Liu, H. (2016). Performance analysis of a parabolic trough solar collector using Al2O3/synthetic oil nanofluid. *Applied Thermal Engineering, 107*, 469–478.

Wei, X. D., Lu, Z. W., Wang, Z. F., Yu, W. X., Zhang, H. X., & Yao, Z. H. (2010). A new method for the design of the heliostat field layout for solar tower power plant. *Renewable Energy, 35*, 1970–1975.

Wu, Z., Li, S., Yuan, G., Lei, D., & Wang, Z. (2014). Three-dimensional numerical study of heat transfer characteristics of parabolic trough receiver. *Applied Energy, 113*, 902–911.

Xiangtao, G., Fuqiang, W., Haiyan, W., Jianyu, T., Qingzhi, L., & Huaizhi, H. (2017). Heat transfer enhancement analysis of tube receiver for parabolic trough solar collector with pin fin arrays inserting. *Solar Energy, 144*, 185–202.

Xu, C., Chen, Z., Li, M., Zhang, P., Ji, X., & Luo, X. (2014). Research on the compensation of the end loss effect for parabolic trough solar collectors. *Applied Energy, 115*, 128–139.

Yilmaz, I. H. (2018). Optimization of an integral flat plate collector-storage system for domestic solar water heating in Adana. *Anadolu University Journal of Science and Technology A-Applied Sciences and Engineering, 19*, 165–176.

Yilmaz, I. H., & Mwesigye, A. (2018). Modeling, simulation and performance analysis of parabolic trough solar collectors: A comprehensive review. *Applied Energy, 225*, 135–174.

Yilmaz, I. H., & Söylemez, M. S. (2016). Transient simulation of solar assisted wheat cooking by parabolic trough collector. *Journal of Global Engineering Studies, 3*, 93–106.

Zadeh, P. M., Sokhansefat, T., Kasaeian, A. B., Kowsary, F., & Akbarzadeh, A. (2015). Hybrid optimization algorithm for thermal analysis in a solar parabolic trough collector based on nanofluid. *Energy, 82*, 857–864.

Zhang, H. L., Baeyens, J., Degreve, J., & Caceres, G. (2013). Concentrated solar power plants: A review and design methodology. *Renewable and Sustainable Energy Reviews, 22*, 466–481.

ZhangJing, Z., Yang, X., & Yaling, H. (2016). Thermal analysis of a solar parabolic trough receiver tube with porous insert optimized by coupling genetic algorithm and CFD. *SCIENCE CHINA Technological Sciences, 59*(10), 1475–1485.

Zhu, J., Wang, K., Wu, H., Wang, D., Du, J., & Olabi, A. G. (2015). Experimental investigation on the energy and exergy performance of a coiled tube solar receiver. *Applied Energy, 156*, 519–527.

Zhu, X., Zhu, L., & Zhao, J. (2017). Wavy-tape insert designed for managing highly concentrated solar energy on absorber tube of parabolic trough receiver. *Energy, 141*, 1146–1155.

Index

Printed in the United States
by Baker & Taylor Publisher Services